Rescue by Discovery

Hearts of Ice and Steel

Rescue by Discovery

Hearts of Ice and Steel

Beatrice Cowan

Grosvenor House
Publishing Limited

This book is published by
Grosvenor House Publishing Ltd
Link House
140 The Broadway, Tolworth, Surrey, KT6 7HT.
www.grosvenorhousepublishing.co.uk

A CIP record for this book
is available from the British Library

Paperback ISBN 978-1-83615-037-4
Hardback ISBN 978-1-83615-038-1

To my parents, Leonard and Joyce Hill

RRS *Discovery II*

Contents

List of Illustrations

List of Maps

Preface

Among the meticulously recorded details in a programme of groundbreaking scientific research in Antarctic waters known as the Discovery Investigations is a series of reports made regularly by the successive captains of RRS *William Scoresby* and RRS *Discovery II*, describing the overall movements and day–to–day progress of each ship.

One of these, the Forty Ninth Letter of Proceedings of RRS *Discovery II*, covers part of an unusual break in the normal research programme when the ship was sent to the rescue of two American aviators, Lincoln Ellsworth and Herbert Hollick-Kenyon.

Leonard Hill, captain at the time and author of the Forty Ninth Letter of Proceedings, kept his personal copy of this report on a cupboard shelf for most of the following years, and it only came fully to light after his death in 2003. Hill left various notes and jottings which suggest he would have liked to have written an expanded version himself.

The book which follows is an attempt to put the account into its wider context.

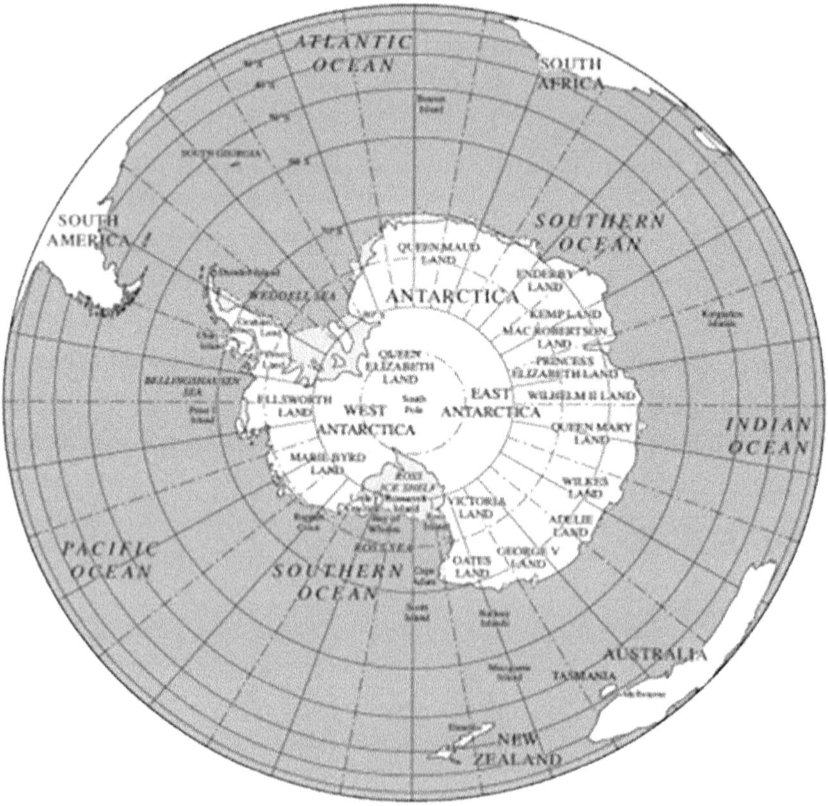

Map 1 Antarctica, with South America, South Africa Australia, and New Zealand.

Introduction

For several months during the winter of 1935–1936, newspapers around the world carried the news of the disappearance, somewhere on the Antarctic continent, of an American explorer and aviator, Lincoln Ellsworth, and his pilot, Hubert Hollick-Kenyon. The flight was remarkable for the way in which it revealed a stretch of land, hitherto unseen and therefore uncharted and unexplored. Unfortunately for the aviators, their radio connections failed, and the waiting world became alarmed. Their rescue six weeks later, resulting from a remarkable union of traditional seamanship and the latest achievements in aerial exploration, caused even more of a stir.

The mid-1930s marked a time of change in Antarctic exploration. The Heroic days were over. The previous 150 years had been a period of great exploration, highlighted in the first part of the twentieth century by the work of Robert Scott and Ernest Shackleton. Countries claimed or annexed areas, and maps of the time clearly showed the demarcation of the land they had discovered. Antarctica appealed as the last of the great areas for them to acquire. Britain, with her widespread Empire, was at the forefront of those wanting to have a firm foothold in that part of the world. At one point the UK hoped to claim the entire continent as part of its empire, but by the 1930s changes had come.

In 1923 New Zealand assumed from the UK the right of sovereignty over the Ross Dependency. This included the Ross Ice Shelf, the Balleny Islands, Scott Island, and other adjacent islands. In 1933 Australia, which had always shown a strong interest in the entire area, accepted control of the territory neighbouring it to the West, and called it King George V Land. Meanwhile, areas of exploration continued. Between 1929 and 1931 the UK, Australia and New Zealand were major explorers in these territories, while

John Rymill from the UK led an important expedition to Graham Land between 1934 and 1937.

However, from the early 1900s, another matter drew people's attention to the South. As a result of declining stocks of whales in Arctic regions, there was now sharp focus on the whaling industry. It was essential to prevent losses in Antarctic waters similar to those in the Northern waters. For one, the Falkland Islands Dependency depended on income from whaling. To maintain this, the British Government established the Interdepartmental Committee for the Dependencies of the Falkland Islands. Known in more popular terms as the Discovery Committee, this body would carry out scientific investigations into the natural resources of the area, their conservation, and aspects of economic development arising from them.

From the 1920s onwards, steady research by the Discovery Committee produced findings of immense importance to science, with results for oceanography from chemical analysis of the oceans, for marine biology and for geology, all published in the Discovery Committee's publication, *Discovery Investigations.* Specially equipped research ships, the RRS *William Scoresby* and RRS *Discovery,* replaced in 1929 by RRS *Discovery II,* covered many thousands of miles of Antarctic waters and were crucial for the findings, made in a series of expeditions known as commissions.

When the world became aware that there was no contact with Ellsworth and Hollick-Kenyon, the Australian Government felt called upon to organise a rescue. Knowing that a UK research ship was working in Antarctic waters, they asked the UK Government to provide help. In London the Colonial Office turned to their Antarctic Committee who, after much debate, called on RRS *Discovery II,* the research ship in question, to provide the help required. She would be equipped with seaplanes along with airmen from the Royal Australian Air Force (RAAF), then make her way to the Bay of Whales beside the Ross Sea. In due course the New Zealand Government was drawn into the support.

For RRS *Discovery II* this meant the loss of the time and the season for carefully planned research. It also meant a journey through the Ross Sea pack ice, a highly risky operation for a ship which was patently inadequately protected for working through such heavy ice. This entailed passing through 500 miles of often dense pack ice which would, for a while, make any movement impossible and which at any moment could have crushed the small, steel ship into pieces. The Master, Leonard Hill, only recently appointed as captain in readiness for the current commission, had, as part of his contract, agreed to comply with regulations which enjoined due care when entering all but the lightest ice. It was incomprehensible to him now that he should be told to undertake such a mission. However, after several enquiries made to the Discovery Committee in London, and much deliberation and debate with his colleagues on board ship, Hill accepted the task and set out with the support and enthusiasm of his ship's company.

This book focuses on this episode in the mid–1930s when the interests of politics and science clashed and RRS *Discovery II* was diverted from her research programme to assist in the rescue of the American aviators. It includes in some detail the journey of the American aviators and their harrowing journey to the American base, Little America, beside the Bay of Whales. It then describes the extraordinary challenge the captain and the mariners of *Discovery II* met when taking the ship through the notorious pack at the entrance to the Ross Sea in order to rescue them, and the efforts by which this was achieved. It shows the outstanding way in which the airmen of the RAAF assisted the project by identifying a route through the ice, and ultimately were able to overfly the area where they found the missing American airmen. It also recalls a way in which international cooperation made it possible for a team to provide aid in times when resources were still limited, and the risks involved far higher than today.

The book retells the story of the rescue in fresh detail. It adds to the account many of the difficulties in the planning and the preparation of the expedition, less well-known at the time, if indeed acknowledged at all. It considers the stresses and conflicts between the different parties who called for the expedition, the apparent lack of support from some quarters given to the captain of the ship, and the varying and respective priorities of different members of the supporting groups.

It looks at the expedition from the angle of the captain, mariners, and scientists of *Discovery II.* It considers their reactions to the mission and the problems they experienced in communicating with their control centre, the Antarctic Commission in London. It looks closely at initial difficulties in Melbourne which at the time added to the challenge for the captain but, when resolved, led to a supportive friendship and helpful advice for the enterprise. It notes the reactions of the Americans involved at the centre of the rescue, as well as those of both the sailors, scientists, and marine staff of *Discovery II.* It also considers views of the personnel of the RAAF who, from their base on *Discovery II,* created new aviation history in their flights over Antarctic waters and the Antarctic continent itself. For these, it acknowledges the accounts of Flight Lieutenant Eric Douglas, both in his official reports and also from the insights he wrote in his informal 'Rough Log', accessed with thanks to the Scott Polar Research Institute (SPRI), Cambridge.

Without the navigational skills of the Captain, the aviation skills of the RAAF, the companionship and support of the Chief Scientist of the Discovery Investigations, and the enthusiasm of the entire group of men, the outcome of their flight might have been very different for the American aviators.

The book starts from the recollections of the captain of RRS *Discovery II,* both formal and informal. It looks at the records of the Discovery Committee as seen in the ship's logbooks, in the captain's regular letters of proceedings, and in the minutes of the many meetings of the Discovery Committee and its subcommittees. It studies letters sent between government officials, considers the newspaper articles written at the time, and examines much of the correspondence, often private, between committee members. It considers the accounts of the chief air officer, Flight Lieutenant Eric Douglas, on whose help the captain depended for knowledge of the route he might take through otherwise intractable ice, and who, on a difficult flight over the area around the Bay of Whales, effected the sighting of the American aviators. The learned articles in *The Geographical Journal* of the Royal Geographical Society of London have provided a valuable source of information. This book also refers to the accounts of some of the people who were on board *Discovery II* at the time of the rescue, to the book *A Camera in Antarctica* by the photographer, Alfred Saunders, and to the book *South Latitude* by the scientist, F.D. Ommanney. It turns also to Lincoln Ellsworth himself for his detailed accounts in *The Geographical Journal* of London, *The National Geographic Magazine* of the National Geographic Society of Washington D.C. and his book *Lost Horizons,* published in 1938. On occasions, accounts recorded in these works differ. I have endeavoured to assess the events as they happened. The error is mine if on occasions I have made a wrong interpretation.

ACKNOWLEDGEMENTS

For the opportunity to write this account, I am grateful to Naomi Boneham, Archivist, to the staff of the Scott Polar Research Institute in Cambridge for access to papers connected with the RRS *Discovery II* expeditions, and to staff of the library of National Oceanographical Institute in Southampton, where the constant, friendly and knowledgeable support of Jane Stephenson and Emma Guest have been invaluable. I owe it especially to Emma for her kindness in providing the images from the National Oceanographic Library Archives (NOL).

I am also indebted to The National Library of Scotland (NLS) in Edinburgh, which holds the remarkable collection of the papers of Sir James Mann Wordie, with their numerous letters to and from many of the members of the Discovery Commission; to Mr Duncan Stewart, Head of the Print and Design Unit, St Andrews University, for the meticulous work in preparing the maps; to Leona Thomas, daughter of Leonard Thomas, for her friendship, for the chance to see some of her father's Journal written on board *Discovery II* throughout the 1930s and to publish comments from it. I would like to recall the happy memory of the warm reunion, 20 years later, of old friends, Dr George Deacon, later Sir George Deacon, by that time founder and Director of the National Institute of Oceanography, and my father, Leonard Hill, Captain of *Discovery II* from 1935–1939.

I would also like to thank Dr Suzanne Dalton in Oxford and Dr Peter Cundill in St Andrews for their generosity in reading the manuscript at various stages and for encouraging me in this project. Last, but not least, I would like to express my deep gratitude to my husband, George Cowan, for his encouragement, critical questioning, and long-term and unfailing support throughout.

Chapter 1 *DISCOVERY II* AT THE ICE EDGE

It was early December 1935, and almost high summer in the Southern Ocean. Days were long, with little darkness and relative warmth. That morning there was light snow, but the water was calm. These were often the best conditions, allowing for the ship to remain steady while the light was diffused through a thin veil of snow, and there was less glare. The scientists on board *Discovery II* had just completed several hours of work. Now they were looking forward to breakfast before starting the rigorous task of recording and collating the information they had gathered.

The RRS *Discovery II*, a research ship with arguably the best specifications of the time for deep sea scientific research, was working near the ice edge at 59° 44′ S 98° 5′ E, off the Queen Mary Coast of Wilkes Land in Eastern Antarctica. She had left London on 3 October, at the start of a cruise in the Southern Ocean. After a reasonable voyage, they had sighted Robben Island, South Africa, soon after 05.00 one month later. The ship then berthed in Simonstown harbour just after 17.30 on 3 November.[1] For the next week the usual harbour routine took over: repairs and organisation to prepare for the next few months, with official visits but a relaxed atmosphere when officers invited the seamen to lunch or tea. The good relationships which already existed between everyone on board became even stronger. Over the next six days they left a good impression in Cape Town. Early on 9 November *Discovery II* began to move from harbour. She rounded Cape Agulhas early the next morning.

The ship then moved south-east, heading towards the ice edge. This was the start of a series of voyages planned to take them round the entire Antarctic continent over the next nine months.[2] Within a week she reached the region of the Kerguelen Islands, rugged outcrops of volcanic rock in the south of the Indian Ocean forming an archipelago and beset by strong winds and a generally chilly climate. Scientists began to make detailed observations of the deep currents in the Indian Ocean, to see how these interacted with those which flowed westwards from further to their south. The study of plankton would continue throughout the expedition. That November the area lived up to its reputation for fierce storms and wild seas. *Discovery II* was forced to remain hove to on several occasions and the scientists achieved little.

Gradually the wind decreased, and at last, on 21 November, the scientists could start their work in earnest. By now they were enjoying daylight for nearly eighteen hours a day. They came to the pack ice in late November. The original plan had been to take a zigzag course to cover as much distance as possible, but by then, because of after the stormy weather they had recently experienced, the ship's fuel supply was becoming lower than they would have liked, so they stayed close to the ice-edge and followed its line eastward.

After the stormy days at the first part of the voyage, the weather was as good as they might have hoped for the work they needed to do. The wind was light, the sea almost calm. The ship rolled gently, and there was time to pause for two periods each day. During these times, 'stations' as they called them, when the ship remained hove to, the scientists would lower nets into the water and take samples of the plankton and the other marine life which they then drew up. Their aim was to record the details not only of the biological and chemical data they found but also of the sea temperature, its salinity and its currents.[3] They would then collate their results in their laboratories below deck, eventually to present them in the papers of the numerous 'Discovery Reports'.

Heavy pack ice blocked their way to the south, but *Discovery II* stayed at its edge and moved in and out of the lighter pack all day, staying for most of the time on about the 59th parallel, over 60 miles from the Antarctic mainland. They were all enjoying the opportunity to get on with the work they intended, scientists helped by mariners, carrying out stations at regular intervals.

RRS *Discovery II* was one of two ships working in the South Atlantic and in Antarctic waters for a Committee of the Colonial

Office, the Discovery Commission, to research conditions in those waters. Many other conclusions would come from their research, but the overall brief was to find ways to maintain the whaling industry which in Arctic regions had all but run out of whale stock by the end of the nineteenth century.

Captain Cook had reported vast numbers of whales in the southern waters over a hundred years before. Mindful of this, the whaling industry moved to the south, led by the Norwegians, whose Captain, Carl Anton Larson, built the first whaling station at Grytviken on the north-east coast of South Georgia in 1904. Whale oil, rendered down from the blubber, had been used extensively for candle wax, soaps and margarine, while the baleen from the bones of baleen whales was used for ladies' stays. By the early twentieth century the advent of kerosene, vegetable oils, and Bakelite had, to a large degree, reduced the demand, but the unquenchable call for other items such as soap, margarine, lubrication and tanning continued unabated. The First World War, from 1914–1918, then led to a demand for glycerine, also extracted from whale oil, to supply the munitions industry. By the end of the war vast numbers of whales had been destroyed in Antarctic waters.

Self-regulation had not occurred within the whaling industry itself, and it looked increasingly as if the same fate would remove the whale stock in the Antarctic, just as it had had already devastated numbers in the north.[4] The Falkland Island Dependencies, created in 1908, had come to rely on income from dues from the industry. It was feared that the whaling industry would soon be at risk and that, as a result, the whole economy of the Falkland Islands be challenged. Moves had begun before the 1914–18 war to investigate the situation further.[5] The overall remit of the Discovery Commission was to help the enquiry into whale movements and their feeding grounds. For this the Commission delivered powerful and valuable findings. Originally intended to enable the protection of whales and so enhance the whaling industry, even in the 1930s, movements developed from their work to preserve whales for their own sake. Though at first considered by some as secondary to polar exploration on land, the focus of the Discovery Commission came to provide indispensable support for work in the Antarctic through the charting and survey work it undertook in addition to its scientific research.

A forerunner of the Discovery Committee was formed in 1918 to explore ways of conserving resources of the seas around the Antarctic. One of the Discovery Committee's primary aims was the scientific regulation of the whale fisheries of the area, but it then inevitably included the study of all the other influences on the whale population, whether biological or oceanographical. Funded by the Colonial Office to which it was answerable, it became the official Discovery Committee in 1923, charged to review the natural resources of the Southern Ocean in all its aspects. In its lifetime the Committee sponsored a series of expeditions to southern waters, to research the biology and hydrography of the entire area. Work began in 1925, and a Marine Station was built at Grytviken in South Georgia as a base for research.[6] These 'commissions', as the expeditions were known, would result in a wide range of findings which were steadily reported to the various subcommittees of the Discovery Committee. Sometimes gradually, sometimes dramatically, they helped to change views both about whaling and about the nature of the southern oceans.[7]

The first ship the Discovery Committee sent to the South was RRS *Discovery*, the wooden ship powered by steam and sail and icon of the Heroic Age of Antarctic exploration, best known for her part in the British National Antarctic Expedition under Captain Scott between 1901 and 1904. The Committee took on a second ship, the RRS *William Scoresby*, in 1925–6. Dependent on sail as well as steam, *William Scoresby* was purpose-built for research and whale-marking. The process of marking consisted of firing a dart, later a small harpoon, described as a long bullet, into the whale's blubber. This was long enough to become sufficiently embedded in the blubber and so able to be found by workers at a flensing station, should it be caught, but not so big as to cause harm. If the whale were subsequently taken, the harpoon would identify its area of origin. Use of the harpoon provided interesting information on the movement of whales. Scientists on *William Scoresby* continued in this work until the start of the Second World War and produced valuable results throughout the time the method was practised.[8]

The original *Discovery*, however, became unsuitable for the kind of study the scientists were beginning to envisage. Though constructed of wood, and therefore highly suitable for work in heavy ice, by the mid-1920s she was proving too slow for the work being planned in more open seas and around the entire Antarctic continent.

Encouraged by Stanley Kemp, its chief scientist at the time, the Committee drew up specifications for a ship which would be both faster and more versatile, and so able to work more efficiently and over a wider area. *Discovery II*, as the new ship was named, was not intended to enter the thick pack ice of the coastal Antarctic regions so could be built of steel. She might not have the strength of the old-style wooden ships whose bows were able to ride up when confronted by heavier ice, but she would offer better accommodation for living, provide far more space for scientific work, and carry equipment considerably superior to what the scientists had previously used. Some whaling ships were built of steel, with strengthening for light pack ice. New whaling ships of the factory variety had shown it was possible for a steel ship to enter light pack ice, provided the hull was reinforced to resist any she was likely to meet. As a result, strengthening for work in light floes was eventually built into the specifications for the new ship. To this end, *Discovery II*'s bow and forefoot were designed not only to assist in her steering but also to help her to break light pack ice with reasonable ease. Among many other important features, the rudder was a particular item chosen for upgrade and was much heavier and larger than would normally be required.[9] This enabled the ship to remain on course until almost stationary, an important point, as she frequently had to remain in one place when scientists worked stations or officers took soundings. *Discovery II* had a cruising range of 8,000 miles at full speed, while at an economic speed she could cruise for 10,500 miles. All aspects of *Discovery II* would be checked and refitted every time she returned to London at the end of a commission, and equipment was regularly upgraded.[10]

When she was launched in late 1929 *Discovery II* held an excellent range of scientific equipment, representing the best of the time. She had on board two laboratories, one for biological work, the other for physical and chemical research, with doors leading directly onto the deck and steel storm doors for protection in stormy weather. She contained newly devised aids for surveying, the latest depth-sounding equipment, recorders, and an anemometer. Living conditions were reasonable for the time, with single cabins for the officers and senior scientists on either side of the ship. The petty officers enjoyed double cabins, the seamen bunks. The wardroom for the officers offered accommodation for up to twenty people. It seemed small when all the officers and scientists met together, and even smaller when they entertained visitors on board. Windows on three sides let sufficient light into the oak-panelled room to relieve the gloom. The wardroom had a small table as well as two dining tables. Small bookshelves, with books carefully chosen for their variety and interest, and a wine locker completed the effects. In spite of its seemingly limited size, the wardroom made an excellent meeting place.

In three commissions between late 1929 and early summer, 1935, the research ship covered many thousands of miles. She was the first to circumnavigate the Antarctic continent and in her first six years had produced major scientific results. The earliest work had focused on the whale population and its distribution. Other important topics soon emerged to cover a wide range of fields, from the location and supply of plankton and crustacea to the movement of whales, and from geology to hydrology. Conscious of the range of temperature between summer and winter, they made careful note of the movement of currents, the salts and gases which were dissolved in the water and much more.[11] By December 1935 more than ten of the thirty-seven volumes of Discovery Reports had been published. Scientists had charted the movement of the ocean currents. Marine staff had surveyed crucial areas of Antarctic waters and contributed widely to Admiralty charts. They made important advances by providing answers for gaps in hydrography and cartography which had remained unfilled from the days of explorers such as Cook and Bellingshausen.[12] The Meteorological Office depended on their ice reports which, when published, were sent to other such offices in Australia and the United States and used by seamen throughout the southern hemisphere.[13] *Discovery II*'s findings exceeded anything anyone had anticipated in the early days of the Committee.[14]

Chapter 2 Fourth Commission

Now fresh from a refit which took place between June and September 1935, after her return from the third commission, *Discovery II* was in the early stages of her fourth expedition to the Antarctic. The refit had been extensive. In addition to the usual repairs, the research ship had two new searchlights fitted, one on each side of the bridge. The gear for recording the depth of water had been upgraded. The original system which relied on Lucas and Kelvin machines was retained, but *Discovery II* now had two echo-sounding machines which met Admiralty requirements for taking soundings: two deep-water sounders and a machine for work in shallow water, all incorporating the latest technology. Both the shortwave and the longwave transmitters were replaced with newer sets, hired from Marconi.[1] The motorboat was repaired extensively. Some of the engine parts were replaced.[2]

The programme for this new expedition had been carefully prepared and was both exciting and challenging. It included the acquisition of further information about the movement of water and of plankton. Hoping to build on the experience of the second commission when they first circumnavigated the Antarctic continent, the scientists planned to get as close to the ice edge as possible and look at the number of whales in each area. In so doing, they intended, during the months of the Antarctic summer to come, to circumnavigate afresh the entire continent to complete that survey, and to follow up other research first done in 1933–1934 on the number of whales at the ice edge.

On leaving Cape Town they would take a line of stations, working first from east to southeast and then southwards to the ice edge south of the Kerguelen Islands. Their aim was to have more precise information about the role played by the deep currents of the Indian Ocean and the part they played in the current which flowed westward from there towards the Weddell Sea. Next, when they worked south of the western coast of Australia, they intended to find detailed information about the current which left the Indian Ocean in that region and could then be traced eastwards as far as the west coast of South America. They planned also to spend some time determining the movement of currents south of Tasmania and New Zealand which flowed there from the Ross Sea and of which they had little knowledge. At the same time, they would check the properties and plankton of the water in these currents. These projects would take them until around March 1936, the end of the summer season in the Antarctic. They would then make their way east and north-east to Port Stanley in the Falkland Islands, and from there move to South Georgia before returning to Cape Town at the end of the southern winter. Only then, before the next southern summer, would they pause and take time for refitting the vessel. Throughout the entire voyage they would be taking lines of stations, adding to their steadily accumulated knowledge.[3]

It was stimulating work, but physically extremely challenging. Scientists, helped by sailors, spent many hours on deck in bitterly cold conditions, very frequently working overnight, straining with nets and heaving on ice-covered ropes to haul up bottles of water samples and meshes of biological specimens from a sea where the temperature was always sub-zero. For the sailors this was a specialised occupation. They were required to handle the ship in water that was frequently ice-infested, holding the vessel hove-to over long periods in order to answer the scientists' carefully considered needs. But the sailors knew their role as support cast of a minutely and scrupulously planned programme. Central to the work was the close cooperation between the chief scientist and the captain.

The chief scientist on this commission was George Deacon. The marine officer with executive command was Leonard Hill. Born in 1906, Deacon had joined the Discovery Commission in 1927 as a graduate in Chemistry from King's College, London, and had served as Hydrologist from that time, sailing first on *William Scoresby*. He then sailed with *Discovery II* from 1930–1933. By the time Deacon became chief scientist for the fourth commission in 1935 he had already identified the Antarctic convergence, the point where cold water from the south met warmer water of the north. His seminal account, *A General Account of the Hydrology of the South Atlantic Ocean,* was published in the Discovery Reports of 1933.[4] This was a groundbreaking study which created the basis for all such future enquiries.

Hill, three years younger, had been appointed captain in August 1935, just four months before.[5] Aged only just 27 when he took command, he was considerably younger than the two previous masters, William Carey and Andrew Nelson. This 'old man', as the captain was affectionately known, was far from old. But, young as he was, he had experienced all weathers and conditions.

He first joined the staff of the Discovery Committee in late 1931.[6] Having sailed in 1931 on *William Scoresby* to the Falkland Islands and from there to South Georgia, he joined *Discovery II* in early 1932. He signed his first entry in *Discovery II*'s logbook when she sailed from Grytviken on the evening of 1 March 1935. Hill soon impressed Captain Carey by his calm and competent manner, by his good relationships with the others, and his ability to lead.[7] He became chief officer in the autumn of 1933 under Captain Nelson. In this role he had all but commanded the ship on her return from South America earlier in 1935 when Nelson was unfit. His experience was well-rounded but, because of his young age and looks, Hill's appointment as commanding officer surprised a number of the men. In the journal he kept throughout his time in *Discovery II*, Leonard Thomas, working below decks as a fireman, notes how at the outset Hill was undoubtedly on approval with the seamen on board ship.[8]

At the time of his appointment no-one questioned Hill's navigational skills, but some of the Discovery Committee queried whether one so young would have the ability to manage the other men on his staff. Not long before there had been tensions on board *Discovery II* between some members of the marine staff.[9] As recently as September 1935, a disgruntled member of the ship's personnel had presented the Discovery Committee with a complaint about some perceived events on board ship earlier in the year. A full enquiry into this had been dismissed as unnecessary and serving no useful purpose. How would Hill, as Master, cope if that sort of thing arose again, this time under his command?

J.O. Borley, the Secretary of the Committee, had written privately to an influential member of the Committee, James Mann Wordie, to express his concerns. Borley went so far as to ask Wordie whether it might be advisable to send a senior naval officer, such as the Hydrographer to the Navy, to be with the ship as far as Cape Town to give her a good start. The Hydrographer, Borley suggested, would offer invaluable advice to the young captain, and subsequently report back to the Committee.[10] Wordie, whom Borley frequently consulted for his measured opinion, dismissed this idea outright. It would show, Wordie replied, insufficient trust in the new captain. Far better to leave it to Hill and Deacon to handle between them any problem that might arise. The Hydrographer might, perhaps, sail with Hill and *Discovery II* for the first twenty-four hours, but for longer would be unwise.[11] In spite of Wordie's reply, a confidential meeting was called to discuss the question, but nothing more was heard on the matter.[12] It was known that Hill commanded the respect of the men, and Committee members who attended that meeting clearly felt they could rely on their new captain to handle clashes of personality. Wordie, however, wrote privately to Hill just before *Discovery II* left St Katharine's Dock in October, asking whether he felt up to the task. Hill took his time to reply, waiting until he had a moment in Cape Town itself. Perhaps he had been unsure what to write, but when eventually he did respond he said he knew Deacon well, they were old friends, and between them they would 'manage fine'.[13] It was clear, even by the time *Discovery II* reached Cape Town, that relationships among the ship's personnel were running smoothly. The mood on board was reflected in the image they projected on shore and they always left a good impression in Cape Town. In the words of the Commander-in-Chief, Africa Station, who had written to Vice Admiral Sir Percy Douglas, of the Discovery Commission, less than a year before:

> I think the 'Discovery II' fellows are doing very well. People at the Cape think highly of them, and I look forward to seeing them again in a few months' time.[14]

There was no sign that their good reputation there changed in any way after their visit in Autumn 1935.

As First Officer on the previous commission, Hill had already made it clear that not only the seamen but also the officers should help the scientists as much as possible.[15] He now asked for some immediate changes, requiring from the start that, during the ordinary routine of the voyage, the ship's officers would assist Deacon three nights a week with the hoist of deep-water bottles. In Hill's view it was important for the officers to realise just how difficult it was to handle the scientific gear, and Deacon clearly appreciated the extra help. Now too, in addition to their normal navigational duties, the officers were to share the echo sounding watches. Hill appointed Richard Walker, the new chief officer, to have the main oversight of this task. The ship's doctor, J.R. Strong, and the laboratory assistant, B.F. McCarthy, would assist him. McCarthy already had a good knowledge of this part of the work and helped Walker considerably in his new duties.[16] Soon the ship settled in well to its regular routine. Everyone adapted to the new hierarchy and Deacon and Hill were content with the share each had as leaders of the expedition, the one as head of scientific work, the other as overall commander of *Discovery II*. Their ability to discuss their movements at every stage was crucial and their relationship was good. The project was progressing well.

For several days they had been skirting the ice edge, moving in and out of ice streams, steadily working their way eastwards. They were taking short zigzags as they went and were coping easily with the smaller chunks of ice, known as growlers and bergy bits. Even the occasional areas of pack ice were causing no problem. Now, on the morning of 4 December, they had reached a point well over 2,000 miles south-west of Australia, and just over 400 miles from the Antarctic mainland. At 59° 44′ S 99° 05′ E, they had reached a parallel at which they had previously worked after leaving South Georgia on earlier commissions. This suited their research admirably. The ship was moving at a reasonable speed between stations. The wind blowing from the south-east was no more than a fresh breeze, the waves were barely lengthening. Only a gentle swell affected the ship's position, and the visibility was good. Conditions for work were proving as comfortable as they could expect. They could ask for little more.

Suddenly a stir ran through the ship. A.E. Morris, the telegraph operator, sitting at the telegraph station in his cabin on the main deck, blinked his eyes in surprise as he listened to his shortwave radio. It was not for him to speak of what he read, so he simply sent the message to the bridge where the captain and the chief scientist were talking of the day ahead. Hill looked at the message in silence, then handed it to Deacon:

> Practical means of cooperating in search for Lincoln Ellsworth and colleague are under consideration. You should proceed with all possible despatch to Melbourne. Further instructions will follow. Report noon position daily. Secretary.[17]

'Abandon Stations and make for Melbourne' they each registered in puzzled silence, 'with the utmost despatch.' They gasped in dismay. They were pleased with the way work was going. They had not expected, and certainly did not want, an interruption to its progress. While Hill composed his reply to the Discovery Committee, Deacon went quickly down to the wardroom situated on the level below, on the level of the main deck, where scientists and those officers who were not on duty were at breakfast. They too sat back in surprise. 'We're going to Melbourne. Right now!' Deacon said, waving the telegram in the air. 'Top speed!'

Deacon, normally always calm and measured, spoke more crisply than usual. He, above all, knew how well their research had been going. So much at the mercy of the weather conditions, they were currently able to move quite freely in the region they were examining. The ice was relatively light, and the few streams of it they did meet could be nudged aside gently.[18] Results were proving highly satisfactory.

The telegram gave no further reason for this sudden recall, other than co-operation in a search. Clearly, they were to assist in the search for Lincoln Ellsworth, the well-known American explorer. Radio chatter had already revealed that he might be in difficulties. When people needed help, there would be a natural, human response for those in the Antarctic to help each other where possible. *Discovery II* had taken relief to an American explorer, Admiral Richard Byrd, in 1933.[19] She had been on standby earlier in 1935 to help Ellsworth when he had encountered problems on a previous expedition. This was another call for help. What they could do by going to Melbourne was another matter.

Hill was in no position to question the full reason at this point, and when, at midday, Morris tapped out his reply to the Committee in London, the telegram simply read:

> 1200 4th December R.R.S. Discovery II Latitude 59° 3′ South Longitude 99° East. Your telegram no. 20 proceeding with all despatch to Melbourne. Expect to arrive 15th December. Hill.[20]

Deacon added his own reply. Not all of this reached London clearly, but his message was similar, saying that he would relinquish all stations until he heard to the contrary.

Chapter 3 FULL SPEED TO MELBOURNE

Reactions on board *Discovery II* were varied. First was the question of mail. Christmas was not far ahead. From such a distance in their hemisphere this was even more important a time to remember and, in some way or another, to be in touch with family and friends at home. They would now have to forgo their stay in Fremantle, where they had been due on 18 December. As a result, they would miss the chance to pick up their Christmas mail there and be unable to send their own correspondence home to arrive in time for the festival. They also had with them mail for the people of an expedition being led by John Rymill, which was at that time working in Graham Land.[1] Up till then the plan was for *Discovery II* to arrive at Port Stanley in March-April, then move south to the area around Graham Land. It was now doubtful when they would reach that point in their schedule. In Graham Land itself, Rymill and his colleagues would have a long wait for their postbags to arrive. Then there were *Discovery II*'s own Christmas celebrations. They had planned to remain in Fremantle until 26 December but would now have to forego a Christmas ashore and accept at sea whatever conditions the weather brought for the festive season.

There would, however, be compensations. Some welcomed the thought of being able to enjoy warmer weather a few days earlier than they had expected. Others were already glad of the break from the exhaustion of hauling the nets and the overnight stations. For those with family and friends in Melbourne, whom they had been due to visit after leaving Fremantle, the arrival nearly three weeks earlier than planned would be pleasant. They were originally expected to arrive in Dunedin, New Zealand, at the end of January and spend three days there before moving on. They hoped that they would still be able to go there, perhaps for further supplies, before finally turning for the south.

The chief engineer set the engines to run at around 100 revolutions per minute. For the next few days they averaged a speed of nine to ten miles an hour as they ran north-eastwards to Melbourne. To facilitate this Hill ordered the small mizzen sail to be set.[2] A wind from the south-south-west favoured them, often blowing a fresh to strong breeze, force 5-6, with large waves forming and spray covering the foredeck. The mizzen added noticeably to their speed. It saved some oil and made for steadier and more pleasant sailing. Once they had removed the sail from its locker, the crew took advantage of the moment. It was a rare event to have the sail out and in use, and it brought added satisfaction all round as they cleared the locker and sorted its contents.[3] On 8 December, the engine ran between 103 and 105 rpm. and they covered up to ten miles per hour. At this point the conditions were so much in their favour that they could afford to use more fuel. So far, they had moved forward well as they sailed away from the ice edge. Now Hill told the chief engineer to increase speed to a steady 105 revolutions, and they covered no less than 11 miles per hour.[4]

By 10 December everyone in *Discovery II* had become accustomed to the speed as the ship churned her way through the water. She rolled easily from side to side and barely ever pitched forward into the waves. As the temperature increased, people began to relish the warmth. It gave a welcome reprieve after the days when ice formed on their gloves and on their outer garments as they hauled in nets and speed logs at the ice edge.

A slightly more complete outline of the task ahead arrived two days later, in a different telegram. Sent on 6 December this came not from London but from Captain John King Davis, the Director of Navigation in Melbourne:

> On assumption your vessel proceeds to Bay of Whales after refitting Melbourne please prepare list of engine and hull, overhaul necessary if any on arrival also list of additional stores necessary for four months voyage. Radio your full fuel capacity and estimated quantity remaining on arrival Melbourne also consumption per day at ordinary sea speed.[5]

To be told that their goal was to be the Bay of Whales came as a total surprise to Hill and Deacon. An inlet formed by a gap in the ice of the Ross Ice Barrier, now known as the Ross Ice Shelf, and lying as far south as 78° 5′ S, this was a place well beyond their normal reach. It lay to the southeast of the Ross Sea, itself well-guarded by more than 500 miles of pack ice. *Discovery II* was hardly built for such a challenging passage. As far as *Discovery II* was concerned Hill and Deacon felt sure they could do nothing to help. But the very unlikelihood of their being sent there struck some note of reality in view of the Committee's terse telegram. The Director of Navigation must surely have acted with some knowledge as well as authority.

Hill's reply was friendly but cautious. He admitted that he doubted his ship's suitability and her fuel capacity for such ice navigation as would be required in a search for Ellsworth. However, he sent Davis precise details of *Discovery II*'s fuel capacity when tanks were full, and he gave an estimate of the amount of fuel they expected to have after their journey to Melbourne. A fuel tank could carry up to 316 tons; they used around 8 tons per day and expected to have about 40 tons left when they arrived in Melbourne.

Later that day, 6 December, another cable arrived, again from Melbourne, this time from the Editor of *The Melbourne Herald*, asking Hill for his own views about such an expedition and for the attitude of the ship's personnel towards it:

> Would appreciate short message your view of prospects of Ellsworth search and general attitude of ship towards new expedition.[6]

The mariners had already gleaned from radio chatter in the last few days the news that Lincoln Ellsworth was in some sort of trouble and knew from the initial telegram that it was he who was the cause for their recall to Australia. But this was the first specific reference to anything other than 'co-operation' in a mission to help the American explorer. Hill's reply was straightforward:

> Know nothing as yet of Ellsworth's difficulties very anxious to do all we can.[7]

Soon everyone on board had pieced together at least some small part of the story: the explorer and aviator, Ellsworth, had indeed experienced some mishap. In so far as they knew anything of the American's route, they thought it ran along the west coast of Antarctica, too far away from their current position south of the Kerguelen Islands for them to be of much use. But now, the Bay of Whales and the Ross Sea made a little more sense as an area where the Americans might be in trouble. If Ellsworth and his pilot had ditched in the Ross Sea itself, where water would not be much above freezing point, survival would have been difficult. Alternatively, he and his pilot might have crash landed near to the Ross Sea, presumably somewhere near to the Bay of Whales. So, when were they last seen? How was this known? Questions came thick and fast.

Discovery II continued to push on as fast as possible. Fortunately, the weather on their way to Melbourne remained exceptionally good for such high latitudes. The winds were following and moderate, the weather fine and clear throughout.

The Director of Navigation sent fuller details from Melbourne on 10 December. He assumed that Hill had heard of the change of their plans from the Discovery Committee in London, that he knew their final destination was the Bay of Whales, and that their goal was to rescue the American aviator Lincoln Ellsworth and his pilot Herbert Hollick-Kenyon. These two had last been heard of on 23 November, their position at 79° S 76° W, on a flight from Dundee Island to the Bay of Whales. First of all, Davis wanted to know whether Hill felt confident that *Discovery*, as he called the ship, was up to the task and could safely be deployed on this mission. Everything else depended on that. Hill, Davis wrote, should ignore the fact that there were indeed steel whaling factory ships at work in the area which could make their way through dense pack ice. Factory ships went into the pack of their own free will.

They were accompanied by catchers and took their time before making any decision whether to move forward or to remain wherever they might be.

It was, continued Davis, for Hill himself to decide. The Commonwealth Government of Australia would accept his decision. No pressure would be put on him to go if he deemed it unsafe and an unjustifiable risk for *Discovery* and those on board. Hill must not feel the Australian government's aim to rescue Ellsworth should override his own judgment and lead to any risk that he, as an experienced seaman and, above all, as Master of his ship, considered unjustifiable. Davis' view was that Hill's decision had to be final. Hill, as the seaman, should think carefully through the matter. Davis would appreciate as early a reply as possible once Hill had considered what was involved.[8]

John King Davis was a distinguished sailor and navigator with considerable knowledge of Antarctic waters. The area of the sea they had just left was named for him. He had a reputation for straight talking and, seemingly, had lost friends as result.[9] Yet his telegram appeared thoughtfully worded. What Davis did not appreciate at this time was that no-one on board *Discovery II*, not even the Captain, as Master of the ship and the one who held executive command, could alone make a strategic decision as momentous as one that would take them through the notorious pack ice that guarded the Ross Sea. Such a judgment had to be made in conjunction with the Committee back in London.

So far, Hill had received no further details from the Committee in London after their initial, brief instructions in the first telegram. Clearly, whatever their speculations, the background was far more complicated than Hill or Deacon had imagined. Hill thanked Davis for his telegram. He indicated that, as yet, he had received no instructions from the Discovery Committee other than to proceed to Melbourne as quickly as possible. He certainly suspected that there was a possibility of their helping in their search for Ellsworth but doubted whether *Discovery II* could be used for the kind of navigation through ice that would be necessary to reach the Ross Sea.

Discovery II would also, Hill added, be seriously constrained by the limitations of her fuel supply. Hill asked Davis if he could arrange for a whaler to be available to fuel *Discovery II* in the Ross Sea at the ice edge, both on the outward journey and on their return, as they entered and finally left the ice.[10]

Hill knew of Davis' reputation for straight talking.[11] He was grateful for the considerate telegram. He ended his reply by saying that he wanted to be as helpful as possible and that he welcomed a meeting with Melbourne's Director of Navigation on his ship's arrival in Australia, once he had heard more from the Discovery Committee. At this stage he had no idea just how challenging that meeting would prove.

A second telegram eventually came from the Discovery Committee, in the early evening of 12 December.[12] Now, after their days of waiting, Hill and Deacon knew definitely what they had come to realise over the last six days. The Committee, they read, was anxious to cooperate fully with the Commonwealth Government and to launch a rescue to search for Lincoln Ellsworth and his companion. Once in Melbourne, Hill should arrange a meeting to work out the best way of shipping aircraft and flying personnel to the Bay of Whales. If there had to be any major structural alterations to enable *Discovery II* to carry the aircraft, he should report to the Committee in London and await the Committee's decision. Hill should report the length of time it would take for such alterations. He should also let them know at what point aircraft could be taken on board and the likely date *Discovery II* would be ready to sail. As soon as possible Hill should go as far as feasible through the ice towards the Bay of Whales and should listen for wireless communication from Ellsworth. Hill should stick to the normal standing directions given to the skipper when he took command and navigate through ice accordingly.

Hill was to keep the Discovery Committee fully up to date at every stage, so that it could send whatever instructions necessary. He must also submit the names of the flying personnel who would come aboard ship. The Committee passed on details of the frequencies of Ellsworth's wireless, for which telegraph operators would be listening, and the time scheduled for transmission.[13] They also gave the call sign for Ellsworth's support vessel, *Wyatt Earp,* commanded by Sir Hubert Wilkins. Wilkins would continue to try to communicate with Ellsworth and would make contact with *Discovery II* when *Wyatt Earp* came within calling distance of them. To help them to estimate where this point might be, the telegram stated that the last message Wilkins had received showed that, at the time of transmission, Ellsworth's aeroplane was at 75° S 79° W. It was clear that this was a point beyond the Stefansson Strait. Over what sort of terrain they were flying at the time was quite unknown.

Such was the detail written in this telegram that it seemed to Deacon and Hill that these instructions were final. They were certainly not, at this stage, to be questioned or challenged. As ordered, Hill immediately acknowledged that he had received the details. He then turned to the Chief Scientist, Deacon, the Chief Engineer, W.A. Horton and the First Lieutenant, Richard Walker and began what would seem endless discussions about the enterprise.

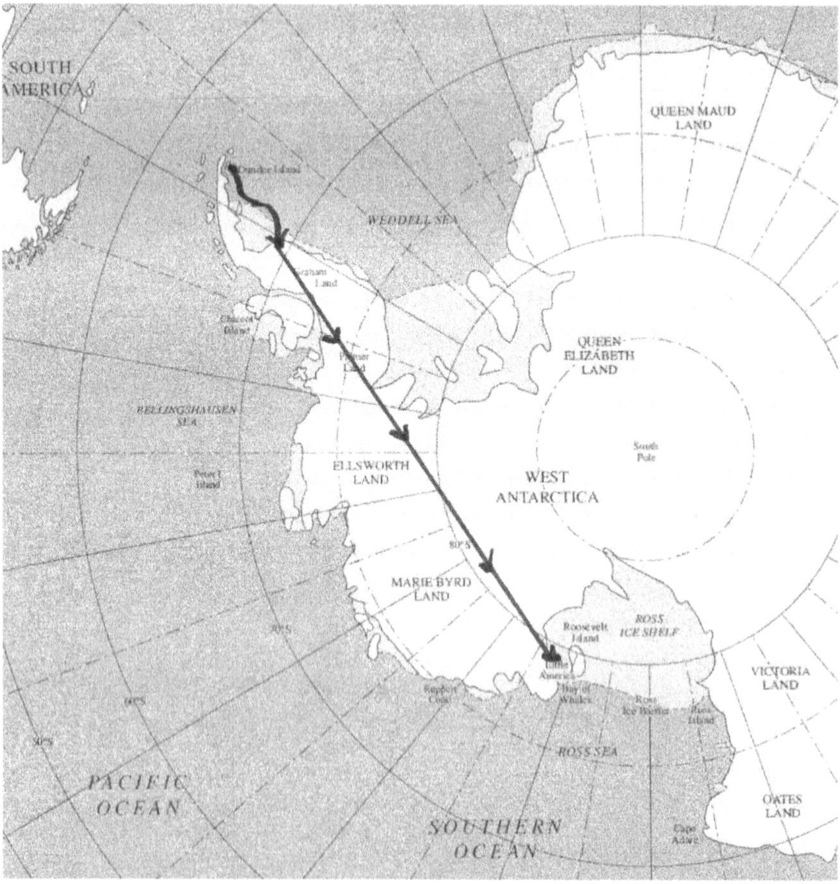

Map 2 West Antarctica, from Dundee Island to Cape Adare, showing route of Lincoln Ellsworth's flight from Dundee Island to the Ross Ice Shelf.

Chapter 4 Lincoln Ellsworth

It was not the first time Lincoln Ellsworth had tried to cross Antarctica by air. Son of a wealthy American and 'a fossil hunter turned explorer', as he called himself, Ellsworth had for some time been thrilled by explorations in the Arctic. More recently he had turned his interest to the Antarctic, where exploration by air was developing rapidly and had in recent years become an important aid to work in that area. Air reconnaissance had first been fully recognised less than ten years before, when Sir Hubert Wilkins was among the first to fly above the continent. Australian by birth, naturalised British, war photographer, aviator, and explorer, Wilkins had already flown in the Arctic when, in April 1928, he made a flight from Alaska to Spitzbergen with his pilot, Carl Ben Eielson, himself an experienced flier. They landed on the ice 600 miles north-west of Point Barrow to sound the depth of the ocean, then successfully took off once more. Their achievement was considered outstanding, both for piloting and navigation.[1]

In the Antarctic summer of 1928–1929, using Deception Island as a base, Wilkins flew along the coast of Graham Land, better known today as the Antarctic Peninsula, in the Pacific and American quadrants of Antarctica. He made an initial flight on 16 November 1928, then repeated the trip a month later to the southwest, flying over much of the same area. A third flight made in January 1929 confirmed to him what he already believed he had discovered: that the Peninsula was in fact an archipelago, separated from the mainland by a strip of water. Wilkins named this the Stefansson Strait. Another channel further north, the Crane Channel, separated the peninsula into North and South Graham Land. He also made what were at the time accepted as further important sightings which received considerable acclaim at the time.

Wilkins had experienced difficult conditions on many occasions. He knew of the need for good radio communications and the importance of avoiding cloud cover when coming to land, but he was fully aware of the sheer thrill of seeing hitherto uncharted area of Antarctica unfold beneath him. Some of Wilkins' findings were later disproved, but his flights showed the great possibilities that aerial reconnaissance opened up for Antarctic exploration.[2]

At the end of 1929, Wilkins returned, now with the support of the British Colonial Office and the Discovery Committee who voted £10,000 for the project and allowed him the use of the RRS *William Scoresby* as a floating base. This time Wilkins' party had much more equipment, including a caterpillar tractor, a boat with an outboard motor, and an eight-wheel vehicle equipped with chains for travel on the snow. Conditions were difficult for much of the time, but on 29 December Wilkins realised that one area of land, hitherto known as Charcot Land, was in fact an island. He had been granted authority to take possession of territory between the Falkland Island dependencies and the Ross Dependency, and named the area Charcot Island in honour of King George V.[3]

In the same year, 1929, the American, Richard Byrd, set up an extensive camp at 78° 12′ S 162° 12′ W near the Bay of Whales. It was this site, at the head of a slope above the Bay, which instantly became known as Little America.[4] With two others, Byrd flew from there to the South Pole and back.[5] He also flew eastward over land which he named for his wife, Marie Byrd Land. Between 1930 and 1931, the Norwegian Hjalmar Riiser Larsen, leader of the *Norvegia* expeditions and already a distinguished aviator, flew several times along the coast to the east of the Weddell Sea. In the same year Sir Douglas Mawson, the Australian leader of the British Australian and New Zealand Antarctic Research Expedition (BANZARE), flew from the research ship RRS *Discovery* and covered over 2,500 miles of the Antarctic coast in the Australian and African quadrants.[6] Both the Swedish and the BANZARE expeditions established fully the value for reconnaissance and survey work, but most of the interior of Antarctica remained entirely unknown. The continent offered seemingly unlimited scope for exploration. Lincoln Ellsworth picked on this territory with enthusiasm.

Ellsworth had already flown in the Arctic, as companion to Roald Amundsen. Knowledge of the recent successes in the Antarctic inspired him to carry out his own aerial reconnaissance. He had met Hubert Wilkins in 1930 and the two had become firm friends. Together they now planned their own expedition to the South. Ellsworth wanted to establish whether, as Wilkins already suspected, the mountains that ran along a line southward through Graham Land were in fact an extension of the Andes Mountains which re-emerged after a dip under the Drake Passage to the south of Cape Horn.[7]

Wilkins became leader of the expedition and in this capacity would control proceedings from the ship. Ellsworth would overfly the area. As pilot, Ellsworth chose Bernt Balchen, a Norwegian by birth, who had previously amassed considerable experience in the Arctic. On 19 November 1929 he had flown to the South Pole as pilot for Richard Byrd. Byrd praised his skill unreservedly. To Ellsworth, Balchen seemed the ideal choice for his expedition.

With Balchen as advisor, Ellsworth acquired an aeroplane, an early Northrup Gamma. Made to Ellsworth's design and specification, this became the prototype for many commercial and military planes to be made by the Northrup Corporation. It had a Wasp 600 hp engine and could reach a speed of 230 mph and cruise for up to 7,000 miles. It was the best of its time for the purpose for which it was chosen. Wing flaps would slow down speed when landing and skis could also be fitted if required. In addition, low wings made this aeroplane ideally suited for Antarctic work, as the fliers could scoop out trenches for the plane's skis once they had landed. The wings would thus rest flat on the soft snow and so prevent the plane from being lifted by the inevitable wind. Ellsworth named the aeroplane the *Polar Star*.

Ellsworth then began to explore the best means of transport. He had heard excellent reports of Norwegian fishing vessels and bought a single-decked, motor-driven herring boat, constructed of oak and Norwegian pine. This had been built in Norway in 1919 as a fishing vessel for use in the Bay of Biscay, and Ellsworth reconditioned her for use in ice-infested waters. He named her *Wyatt Earp*, for his hero, a United States frontiersman and Sheriff of the latter part of the nineteenth century.

The first time they attempted the flight to the Bay of Whales, the party set out in *Wyatt Earp* from Dunedin, New Zealand. After a month they reached the Bay of Whales, an enclave in the ice of the Ross Ice Barrier, and moored at the mouth of the bay. Ellsworth and Balchen winched the *Polar Star* onto the Barrier and made a promising test flight, then left the plane well inland. Disastrously, the mass of ice on which the *Polar Star* rested broke from the Barrier overnight. The *Polar Star* fell into a crevasse and sustained serious damage. On the second attempt they started from Deception Island in West Antarctica, planning to fly southwards following a great circle route. They eventually chose Snow Hill Island in the north-west of the Weddell Sea as their point for take-off. This time Ellsworth and Balchen were thwarted by poor weather at every stage of their journey. It was early January and the flying season well advanced, but renewed bad weather set in and they soon turned back.

No doubt seriously disappointed, Ellsworth resolved to make a third attempt and renewed his efforts to achieve his goal, flying southwards over Graham Land. Ellsworth still felt considerable respect for Balchen but admitted that he felt thwarted by the pilot's cautious approach to the flying conditions. When Balchen moved to commercial aviation in Norway, Ellsworth advertised in a number of aero clubs and aviation companies for a new aviator. He chose two of the applicants. One was Herbert Hollick-Kenyon, a Canadian who had been born in London and had emigrated with his parents as a boy. Hollick-Kenyon later returned from Canada to the United Kingdom and spent some years, initially in the Royal Flying Corps during the First World War and after that in the Royal Air Force, as the Royal Flying Corps became known. He then spent some years flying in the Canadian Arctic under snow conditions before returning to Canada to work for Canadian Airways. The other applicant chosen, J. H. Lymburner, from Ontario, worked as a pilot with the same company. Both Hollick-Kenyon and Lymburner had good experience and excellent qualifications.

Ellsworth set off in the autumn of 1935, in the converted Norwegian fishing vessel, *Wyatt Earp*. With him were the aviators, Hollick-Kenyon and Lymburner, while Hubert Wilkins continued as overall leader of the support team. The route Ellsworth planned to take was the same as he had intended for the second expedition. Once again, he aimed to follow the line over Western Antarctica between the Weddell Sea, south-east of Cape Horn, and the Ross Sea, which lies at almost 80° S and some 3,000 miles to the south of New Zealand. Ellsworth and his pilot would fly over Graham Land, then along the west coast of Antarctica, over Charcot Island and Marie Byrd Land, between 158° West and 103° West. Their final destination was to be the Bay of Whales. This point would provide an ideal place for them to meet Wilkins and the support party, who would arrive by sea in *Wyatt Earp*.

This was Ellsworth's long-intended route, but another reason for choosing this line, which he called 'even more cogent', was Richard Byrd's base at Little America. This settlement, set just four and a half miles from the Bay of Whales and the edge of the Ross Ice Barrier, had only recently been abandoned after Byrd's second expedition. Ellsworth had laid down 500 gallons of gasoline there the year before, along with other necessities, and he understood that there would also be food stored which they could use. He did not know to how much this would amount, but he was well aware that, once at the Bay of Whales, seals and penguins would abound from which they could top up their own provisions. All combined, that would sustain them for at least one month were *Wyatt Earp* to be delayed in reaching them. They would be able to live, Ellsworth wrote, until the end of January or later. Explorer and pilot would have no problem in waiting at the Bay of Whales until Wilkins reached them in *Wyatt Earp*. The greater part of the land over which they planned to fly had not been explored and was, as yet, unclaimed territory. Ellsworth intended to raise the 'Stars and Stripes' there and so claim for the United States the area up to the 120th meridian between Marie Byrd Land and the Bellingshausen Sea.[8] He would name the area after his father, James W. Ellsworth.[9] In taking this route of over 2,140 miles of unexplored Antarctic, he hoped to prove the theory that the Antarctic did indeed stretch as one vast continent stretching from the Weddell Sea to the Ross Sea.

As pilot for the expedition, Ellsworth chose Hollick-Kenyon, now frequently known simply as Kenyon. For their starting place they chose Dundee Island, just south of Joinville Island, at the northernmost point of Graham Land and the north-western point of the Weddell Sea. On their first attempt on 21 November 1935, an oil leak forced them to return after only a few hours. The next day, 22 November, they took off in perfect weather at 08.00 GMT, expecting to take around fourteen hours to bridge Antarctica for the first time. They would let those at base know by wireless how far they had travelled. They would have with them good means of communication. In the air they would use *Polar Star's* own radio, capable of transmitting on any wave of the band between 20 and 80 metres. For use on the ground, they could link this to a portable 300-watt generator which they would charge with a small petrol engine. They also carried with them a smaller transmitter for emergencies which they could mount on a sledge. It seemed reasonable to assume that at least one of these methods would keep them in touch with the support party on *Wyatt Earp*.

Chapter 5 SILENCE

Ellsworth's flight proved remarkable in many ways. He and Kenyon crossed the Stefansson Strait not long after midday. They confirmed that Wilkins' earlier claim that such a division of peninsula and mainland existed but found it considerably narrower than previously reported. Soon after, he named a long range of peaks the Eternity Range; three of them he called Mount Faith, Mount Hope and Mount Charity.[1] Then he and Kenyon realised they were flying over unknown territory, rugged and marked by several mountain ranges. Comprising about 350,000 square miles, this was the land Ellsworth would claim for his country and where, later that day, he raised the Stars and Stripes of the US on the land below. Lying between 80° S and 120° W, to the south of Palmer Land, to the west of the Bellingshausen Sea, and to the north of Marie Byrd Land, this was the area he named for his father, James W. Ellsworth.[2] Soon after, around 19.20, though the visibility was by then becoming poor, he was able to see a long range to the west, stretching for around 75 miles and with peaks in the centre which rose to 13,000 feet. He named this the Sentinel Range. The central peak he named for his wife, Mount Mary Louise Ulmer.[3]

The snow plateau they then flew over, to the north-east of the newly named Ellsworth Land, consisted of endless sastrugi, the troughs and ridges running like long waves across the windswept snow fields. Everywhere were isolated nunataks, peaks of rock projecting though the glacier ice and frozen snow. Though he looked carefully, Ellsworth could see no signs of dangerous crevasses. The land beneath them lay at 4,000 to 6,000 ft above sea level, between the northern portion of what would later be known as the Ellsworth Mountains to their east, and Marie Byrd Land to their southwest. This vast area Ellsworth later named

Hollick-Kenyon Plateau for his pilot, who had been flying steadily all day, advising and commenting as they went.

For most of the day visibility had been remarkably clear. They had flown for nearly 14 hours and, because they had been deflected by the wind and clouds earlier in the day, covered considerably more than the 1,450 miles from Dundee Island. Now, at 21.55, with visibility noticeably diminished, they landed and took sextant observations before making camp for the night. This first camp was situated at 79° 12′ S 104° 10′ W, its height 6,400 feet above sea level.

After a short sleep the fliers took off again the next day at 17.03, not long before noon local time, but they soon realised that visibility was too poor for them to continue. Snow was making it unsafe to fly, and it would have made little point if they could not take satisfactory observations. They landed again after only half an hour and remained in that place for three days, mainly confined to their tent because of a heavy blizzard. They then attempted another flight but were forced to land yet again soon after. In their third camp, bound by ice and snow, they remained for a total of seven days. After a further attempt and another, shorter landing, they finally flew over the last of the plateau on 5 December and saw below them the snow fields of the Ross Ice Barrier, 980 feet above sea level. They now had little over 100 miles to go before they reached Little America.

All seemed set for a landing at the old American base, but after so many take-offs and short flights, as well as the extra mileage they had covered due to side winds, *Polar Star* ran out of fuel. With the Ross Ice Barrier within sight and beyond it a darkening of the sky that suggested water ahead, they came to ground near the northern point of Roosevelt Island, around 16 miles from Byrd's settlement.[4]

Chapter 6 RESCUE ALERT

From mid-afternoon on 22 November, their first day, Ellsworth and Kenyon had failed to make radio contact with *Wyatt Earp*. The time was then 16.15. Ellsworth and Kenyon still had 950 miles to cover if they were to reach their destination. It eventually emerged that blame lay with a faulty switch on the trailing antenna. At the time, however, they could not establish the cause of the problem, and Ellsworth encouraged Kenyon to continue on their course until they reached 80° West, the area Ellsworth hoped to name for his father.

The radio officer on *Wyatt Earp* had for some time complained that reception was poor. Now the issue was critical. The wireless operator tried continuously to pick up some communication at the appointed hour, but his efforts were to no avail. By 26 November, he realised that any attempt he might now make to contact *Polar Star* was futile. Wilkins left drums of gasoil at Dundee Island along with a message, in case Ellsworth and Kenyon should return later and land there, then he ordered *Wyatt Earp* to sail for Deception Island. In doing so, however, Wilkins disregarded Ellsworth's own instructions, written carefully and in detail, stating what the support party should do were they to suspect difficulties.

In these meticulously prepared contingency plans, Ellsworth had tried to make clear that, if contact were lost, the support party should let a few days lapse, then start to establish depots at several points as near as possible to the route he and Kenyon intended to take. Wilkins should place the first depot on Charcot Island, at 69° 45′ S 75° 15′ W, off the south-western point of Graham Land. The next two depots were to be first in Marie Byrd Land, at Mount Mabelle Ridley, the next at Mount Grace McKinley. Depending on where they had landed, Ellsworth and Kenyon would make for one of these places if they were unable to make it as far as Little

America. The explorers would in any case have adequate supplies on their plane to take with them and with which they could sustain themselves for five weeks. In addition, Ellsworth had drawn up what he felt were the precisely calculated hourly positions they would reach in the first 11 hours after they had crossed the Stefansson Strait.

Capable of speeds up to seven to eight miles per hour and with a cruising distance of as much as 11,000 miles, *Wyatt Earp* would reach the Bay of Whales, four and a half miles from Little America, by mid-January. Ellsworth was anxious for the expedition to be self-sufficient and unwilling to have to call on others for assistance. As little fuss as possible should be made if things went wrong. He felt that his preparations were sufficiently meticulous as to prevent any problem arising, whatever might happen to him in the meantime.

But alarm was raised. On 26 November Wilkins issued a report from Dundee Island in which he stated that he and his team were now making preparations to start a search for Ellsworth and Kenyon.[1] Word soon spread around the regions within reach of the radio transmission. Newspapers around the world were already covering Ellsworth's project. *The New York Times* printed the news of his silence immediately, and *The Times* of London followed suit.[2] The Australian Press covered the story from coast to coast.[3] Hopes lingered for a while. New Zealand wireless stations spoke of faint signals reported from one ship in Picton harbour and from another vessel nearing Wellington.[4] But the signals were weak and could not be read.[5] Not long after the news first broke, the Australian Government began to mount a search for the missing explorer and his pilot.

The previous year the US Government had helped the Commonwealth Government of Australia in their search for the aviator, Charles Ulm. Ulm had set out to cross the Pacific but had lost contact in poor conditions somewhere north of Hawaii. The US joined the search, but to with no success. Now the Australians were keen to reciprocate this gesture of help.[6]

On 2 December 1935, the Prime Minister of Australia, Joseph Lyons, submitted a proposal to the British Government, asking that the British Government ship *Discovery*, as the telegram named her, be deployed to help in the search for Ellsworth and Kenyon. The Australians felt that the Bay of Whales was too far from any American base in the Antarctic for Ellsworth's fellow countrymen to be able to help. It would be a friendly offer of aid to the Americans if the governments of Australia and the United Kingdom took action together to rescue the American explorers. They knew that *Discovery* was due in Fremantle in the near future, and suggested that she be despatched to the Bay of Whales to start a search. If possible, they themselves would supply a suitable aeroplane for the British ship to take aboard. The telegram ended by stating that the Australian Government was prepared to share the cost of any action with His Majesty's Government in the United Kingdom.[7] It is not clear whether their omission of the Roman numeral *II* in the name of the research ship was an error of judgment or of typing. It may never be known, from the name they gave the ship at that stage, whether they recognised that she was not the wooden ship *Discovery* in which Davis and Mawson had sailed with the BANZARE expedition, but her successor, the steel ship *Discovery II.*

The world press had been reporting the news of Ellsworth's expedition from its outset and from late November had reported on several occasions about the radio silence from him.[8] Now there was a dramatic increase in the coverage. On 3 December, *The Times* of London reported that the Commonwealth Government of Australia was consulting the Antarctic explorer Sir Douglas Mawson about the search for Mr. Ellsworth, as he knew well the area where it was believed Ellsworth had made a forced landing.

Three days later *The Times* confirmed this and added that that the British Government was in consultation with the New Zealand Government. It also gave details of *Discovery II*'s times of arrival in Melbourne and her proposed departure from there. It gave an outline of her route and details of the equipment the ship would take with her. These initial plans would change several times as the days went by, and from early December newspapers round the world were alive with the continuing developments.

The Australian Government also suggested that wireless messages be broadcast to all whaling ships in the relevant regions of Antarctic waters, asking them to do what they could to help. It soon emerged that the Norwegian Whaling Association had already heard of Wilkins' telegram and had immediately ordered all their floating factories and whalers in the Antarctic to listen night and day for wireless signals from Ellsworth. With 16 factory ships and around 150 whalers in the Antarctic alerted, it should be reasonably easy to receive news from Kenyon and Ellsworth should any be transmitted.

The public was already excited by the story, and newspapers everywhere continued to report developments. Details would be changed and projected timings re-scheduled as Governments advanced their plans. Little escaped the national and international press. One thing was clear. If Ellsworth had intended to keep to himself accounts of his exploration until such time as he succeeded in the venture, he now had no chance of privacy.

Chapter 7 REACTIONS IN LONDON

In London, the Discovery Committee reacted promptly to Joseph Lyons' cabled submission from Canberra and called an Emergency Meeting of their main Committee for the afternoon of 3 December. Members would all have seen press reports of Ellsworth's latest attempt to cross the Antarctic. They would also have read of the most recent concerns about his lack of communication, and all these accounts would certainly have interested them. So far, however, they would not have considered that the silence from *Polar Star* should in any way concern them. Ellsworth had been in difficulties before but had always declined any offer of help. It was likely that he would respond in the same way this time. More relevant on this occasion, Ellsworth's current expedition was on the opposite side of the Antarctic continent from where *Discovery II* was currently carrying out research. That the Discovery Committee might offer help had probably not occurred to its members. Now, however, Joseph Lyons' request elicited a different response.

Normally, members of the Discovery Committee met once a month for their full meeting. Now they broke this pattern and assembled around the long table in the Colonial Office for an extra committee meeting to discuss the emergency.[1] In spite of the hasty summons, not only were most of their regular members present, headed by their chairman, the Earl of Plymouth, but also representatives from other groups who joined them by special invitation. Members of the Polar Committee, representatives of the Board of Trade, the Royal Air Force, and the General Post Office were present. J.R. Scott, the Home Agent for John Rymill's British Graham Land Expedition (BGLE), one well experienced in Polar exploration, also attended. It made a large gathering in the conference room. John Borley, Fisheries Adviser at the Colonial Office since 1928, a long-standing member of the Discovery

Committee, and currently acting as its secretary, was the first to speak.

Borley gave a short summary of events leading up to the surprising news, very much as had appeared in the press that day. The first reaction from the representatives gathered round the table seemed enthusiastic. There was nothing new in helping someone in the Antarctic when they experienced difficulties, and the United Kingdom had offered help on several other occasions. The previous year *Discovery II* had pushed through pack ice to take medical aid to Richard Byrd's ship, *The Bear of Oakland*, then in the Ross Sea when Byrd's own doctor was critically unwell. On that occasion *Discovery II* had left Auckland as late as mid-February. For much of the time thick fog and drifting ice made for difficult navigation, but the research ship made successful contact with *The Bear*, then returned to normal duties so that their scientists could continue their research.[2] Had it been any other time, this might well not have been possible. As records show, only a week later *Discovery II* was in thick pack ice and encrusted with ice and snow. It was usual practice for wireless messages to be broadcast to all whaling ships in the locality. Though often hundreds of miles apart, any ship in the area would naturally offer whatever help possible, especially if it were a medical emergency. Clearly the British Government should do the same and follow the Norwegian initiative in listening for any wireless messages. The room in the Colonial Office buzzed with offers of help and suggestions.

The Committee promptly agreed that, if necessary, they could indeed, as the Commonwealth Government was asking, send *Discovery II* to help. She could be despatched with aeroplanes on board to enable an air search to be mounted as required. The Air Ministry readily offered to consider what assistance they could manage. After the initial enthusiasm, however, the practicalities of such an enterprise soon started to exercise their minds. Firstly, members began to doubt whether *Discovery II* had enough fuel to take her directly to the Bay of Whales, the destination proposed by the Australians. For this she would have to call in at a port in

Australia or New Zealand to re-fuel and stock up with extra supplies. This thought led members of the Committee to begin to speculate how long it would take the ship to reach Fremantle from her current position by the ice edge.

Inevitably, the Committee was concerned about the cost of any expedition. Numerous questions began to arise. The Australian Government had offered to share the cost. A new administration had recently been established in New Zealand: the Committee should find out whether they would be willing to help to finance any rescue. The Committee then asked whether the government of New Zealand was already involved in the search. The rescue was, after all, to be in the Ross Dependency, as the Ross Sea area was known. Under New Zealand's supervision since 1923, this area was, strictly speaking, more the concern of that country rather than that of Australia or even, by that time, of the United Kingdom. Australia accepted control of neighbouring territory which it named King George V Land, in 1933. It had always maintained a strong interest in the entire area with major exploration to their credit. It was recognised that they could well find it of economic value. But the Ross Dependency remained under the control of New Zealand. Their government should undoubtedly be involved.

Next came the question of physical access to the Ross Sea, a vast, shallow area situated some 2,399 miles south of New Zealand, guarded by pack ice which for most of the year was exceptionally heavy. The member of the Discovery Committee considered the one best able to speak on the subject of ice was James Mann Wordie, a geologist. Wordie, at that time Senior Tutor at St John's College, Cambridge, sat on the Discovery Committee as representative of the Royal Geographical Society, of which he was Honorary Secretary. Little more than twenty years before, as a young graduate geologist, Wordie had gone south with Shackleton on *Endurance* and, along with all Shackleton's men from that ship, had spent eight months trapped in the ice of the Weddell Sea before its pack ice finally crushed the wooden ship beyond hope of repair. The men had then camped on an ice floe before making their way to Elephant Island. There, Wordie and the others passed almost five further months on the windswept, ice-besieged island, while Shackleton made his epic sea journey to South Georgia, eventually to return to Elephant Island with the necessary relief for his stranded team. Wordie would always have memories of pack ice and its irresistible force. He was without doubt held in special respect because of his experience.

With such short notice, Wordie had been unable to get to the Emergency Meeting and had phoned John Borley beforehand to present his apologies. The two had spoken at length. Now, at the meeting, Borley raised, on Wordie's behalf, a crucial point which alerted the Committee to the risks involved if *Discovery II* were to go as soon as possible. This was still early December, barely high summer in the Antarctic: the summer thaw was not fully established. With such timing it would not be possible for *Discovery II* immediately to reach the Bay of Whales. According to Wordie, there would be no chance of the ice being free enough for the ship to reach the Ross Sea before the middle of January. At this the Committee members nodded their heads in agreement, as they remembered the practical detail that the safe time to reach the Ross Sea would not start until mid-January. It was generally accepted that the open, safe season would then only last for just one month.

This was a point they felt they knew well but, when stated by James Mann Wordie, it bore especial weight. In his telephone conversation with Borley, Wordie had, however, suggested an alternative. Why not, he had put to Borley, ask for help from John Rymill, leader of the British Graham Land Expedition which was working in Graham Land at the time? Rymill's ship, the *Penola,* was wooden and so probably better suited to the ice than the steel *Discovery II*. However, J.R. Scott, the representative of the BGLE, immediately dismissed the suggestion. *Penola*, once a Breton schooner subsequently converted into a luxury yacht, was only partially strengthened for work in ice.[3] Meanwhile, the range of Mr Rymill's aerplane, as Scott called it, was only 1,000 miles, and even that only when travelling light: the aeroplane only allowed for a crew of two, had no space for passengers or extra food, and the pilot would be unable to go more than 500 miles from his own camp. Of course, Rymill might possibly make a circuit nearby, in case Ellsworth had landed not far away from his starting point and was attempting to trek back. Vice Admiral John Edgell, Hydrographer of the Navy and an influential committee member, remarked that even if Ellsworth had flown as far as 800 miles, as might be deduced from the signals received during the presumed eight hours he and his pilot had been in the air, he would still be nearer Graham Land than the Bay of Whales.

At best, Scott continued, the BGLE might offer their ship, *Penola,* for use as a base. More than that Rymill would be unable to offer. Would it be possible, asked Collins, the representative from the Air Ministry, to use a larger, American plane and fly from Dundee Island itself? The rescue team could perhaps borrow Rymill's petrol.[4]

The meeting ended inconclusively, its tone more negative than they had anticipated, the outcome far from clear-cut. As the chairman summarised, they needed much more information before they could proceed in their plans to support the Australian Government.

Immediately after the meeting Borley sent Wordie a summary of the afternoon's discussion.[5] Officially, he wrote, the Committee felt that if anything useful could be achieved, they would, of course, do their best to help in the search. For the time being, in case she could be of some assistance, *Discovery II* should proceed to Melbourne at full speed. She would reach Melbourne on 15 December. From there she might transport a plane at least some of the way south, as far as she was able. The Committee felt sure that, in the course of a day or so, they would have considered all the options. They would, almost certainly, find that *Discovery II* could not usefully be involved. Added to all this, the New Zealand representative on the Committee was unsure whether his own country would cooperate.

If, by the end of the meeting, there was an open degree of hesitancy, further reservations emerged, off the official record, in the letters which soon passed between some of the members. They appreciated that *Discovery II* be called upon: once in Melbourne, the ship would be nearer to the scene required for action. Importantly, the offer of help had, above all, to be seen as a political move and not one to be made simply because *Discovery II* was, comparatively speaking, in the right area. They would see that the Government of the United States was informed and, at the same time, telegraph Rymill in Graham Land to see what help he might give.

This would nevertheless come at a cost. Scientific work would be compromised, research opportunities missed. The scientists were working to a well-planned schedule, precious time would be lost, and the sequence of their findings interrupted. Even if *Discovery II* eventually managed to return to the same position where the scientists had left off, conditions would have altered. The sequence and timing of their findings was crucial. From season to season, water temperature would be different, ocean currents altered, marine life changed. The interruption would be a serious blow to the entire scientific programme, especially as they had no idea how long the rescue expedition might last.

Considerable dissatisfaction was expressed behind the scenes, yet the official line remained unchanged. The Discovery Committee was answerable to the Colonial Office and must have felt considerable pressure to acquiesce. As a Foreign Office memorandum would note when the US Government expressed its gratitude for the UK's efforts, the Governments of the Empire were naturally anxious to give all the assistance they could. Although they had handed former British territory to Australia and New Zealand, they still felt a deep concern, if not responsibility, for the Antarctic regions.[6] The political line was paramount.

A telegram was drafted immediately to send to *Discovery II*. It reached Deacon and Hill early the next morning, local time, with its order to change their plans and make for Australia as quickly as possible. There was no further explanation. The telegram's arrival marked the moment when the Discovery Committee's research ship began her unexpected journey to Melbourne.

The next day, 4 December, *The Times* published a judiciously worded account of the Discovery Committee's decision. It stated that the British Government was considering as a matter of urgency whether they should help the Australian Government in the search for Mr Lincoln Ellsworth. As a first step, the research ship, *Discovery II* was proceeding as quickly as possible to a convenient port in Australia from which to start their journey of rescue. Considerable discussion would continue.[7]

Chapter 8 Action Stations!

Help or comment now appeared from many directions. In Australia, the Commonwealth Government began consultation with Sir Douglas Mawson, the distinguished Antarctic explorer. Mawson was well acquainted with the area where Ellsworth might be assumed to have to make a forced landing. His advice, the Australians felt, would be invaluable.[1] The Discovery Committee in London swiftly turned to John Troutbeck of the Foreign Office, who had himself been present at the Emergency Meeting. Troutbeck immediately telephoned Hugh Millard, his counterpart in the American Embassy.[2] This move, however, was not only made as a matter of courtesy. It was a crucial way of checking to what extent the United States authorities were themselves taking any steps in a search to find out any practical details and to gain as much information as possible about Ellsworth's movements. Troutbeck followed his call with a letter listing a variety of questions. What, for example, was the latest information of Mr Ellsworth's present position? How much fuel did Ellsworth's aeroplane carry? What was its power and its speed? What were the supplies and equipment he had to hand? Specifically, did he have a tent? Above all, they would find it helpful to know the wavelength and call sign of Ellsworth's wireless apparatus. The list of questions from the Foreign Office was long.[3] The US Government replied almost immediately with thanks for the communication and for the support shown by the British Government. A swift exchange of information soon followed.[4]

Within a few days the Discovery Committee was able to organise a second Emergency Meeting, and on 9 December the enlarged group met for a second time, again in the Colonial Office.[5] Copies of letters from the United States Embassy to the Foreign Office, dated 7 and 9 December, brought committee members and delegates the most recent news, including the latest word from the American Embassy regarding Ellsworth's and Wilkins' call signals, and a guide to Ellsworth's presumed position when news had last come from him.[6] Those present at the meeting also heard details from the US of Ellsworth's resources and of other ways help might be on hand. They also received the latest information from the Commonwealth Government in Australia. They then surveyed other possibilities, looking at the ships already in the area, and considered, in particular, whether Rymill's *Penola* and Wilkins' *Wyatt Earp* could offer help.

In the intervening days the Australians had forged ahead with plans, and the Committee now focused on the aeroplanes being prepared for the expedition. The Commonwealth Government had already procured a Westland Wapiti Mark 1A and a de Havilland Gypsy Moth 60x aircraft as the planes to be shipped in *Discovery II*. These were single-engine bi-planes, float seaplanes equipped for aerial reconnaissance over longer or shorter distances respectively as the occasion might demand. The Australians were providing the aeroplanes with wireless, floats for use at sea, and skis for ice landing—standard equipment for the Australian army. The planes would be fitted with extra tanks to hold the fuel needed for long-range flights and would carry the gear necessary for a search by land. The Australians informed the Committee that they were obtaining information directly from the Captain of the *Discovery II* about the space needed for docking and the general supplies he would require. In this way, the Australians could go into action immediately, if necessary, on the arrival of the ship at Melbourne to prepare her for the voyage south.

The Australians had already selected an airman, Flight Lieutenant Eric Douglas, to oversee flying operations. Flight Lieutenant Douglas had acquired first-hand flying experience in the Antarctic with Sir Douglas Mawson during the BANZARE Expedition on the first *Discovery* in 1929–1931, and was well qualified for the task.[7] He would have, as assistant, Flying Officer A.M. Murdoch. Captain John King Davis, another old Antarctic hand with great experience of the Ross Sea area, was now mentioned. Davis was already advising matters from Melbourne and, at this stage, was proposing that the ship should go via Macquarie Island, direct from Melbourne to the Bay of Whales.

The Australian Government was also planning to set up wireless communication to exchange views with Wilkins, who was established on Ellsworth's base ship, *Wyatt Earp*.[8] The Australians expected *Discovery II*, now named by them The Royal Research Ship *Discovery I*, to arrive in Melbourne on 14 December. She would leave for the Bay of Whales before Christmas. If members of the Discovery Committee were inwardly surprised at the arrangements the Australian Government already had in hand, they appear to have made no comment. For them, the question of the pack ice remained the priority. Would *Discovery II* be able to navigate through it in safety? Wordie, now present at this second Emergency Meeting, stressed again his point that the pack across the entrance to the Ross Sea would make it difficult for *Discovery II* to reach the open water to its south before the middle of January. Borley added that, although strengthened for work in ice, *Discovery II* would not be able to penetrate the pack any further than the factory ships could manage, and these, well reinforced against ice damage though they were, could still sustain considerable damage in that environment. Ice would always be the controlling factor. To leave so early in the season was clearly premature. Unspoken, the feeling may well have been that the Australians were looking too far ahead and moving with too much speed.

The Committee saw no difficulty in having structural alterations made to the ship. They felt they knew every detail of their research ships and considered *Discovery II* well able to take aboard the aeroplanes the Australians were proposing. Once the ship was in Melbourne, the necessary changes could easily be made to give space on deck. The ship's own derrick would then swing the planes aboard. It was, they thought, more relevant to ask whether these aeroplanes really were suitable for the intended task. Would they be capable of sufficient range to be of use? And at what point would alterations be made, if a decision to carry aircraft were only to be made later?

The Committee doubted whether a Wapiti or a Moth would be of much practical use for the purpose. The normal range of a Wapiti was about 600 miles in still air. This might be increased if extra tanks were fitted, as the Commonwealth Government was suggesting. Even then, there remained the question of the manoeuvrability of the two aeroplanes once they were on board. How, the Committee asked, could they be moved from the ship when required for service? A Moth might perhaps be swung out of the ship if they altered a derrick already situated on board, but it would still require a clear stretch of water for taking off. Then, when in the air, its range would be limited. A Wapiti, though capable of covering greater distances and therefore particularly useful over land, might well prove difficult to operate from the ship. In that case, why not take two Moths? asked another, not taking in the full import of the smaller plane's limited range. The debate continued, with painstaking attention to each detail.

For a while the Committee had thought that others might step in to save them from involvement in rescuing the two explorers. They had kept three possibilities in mind, but now these hopes were finally dashed. In the US, the explorer's wife, Mrs Mary Louise Ellsworth, had asked her husband's representative to hire a fresh plane to be shipped out to extend the search.[9] An aeroplane duly left Los Angeles for Punta Arenas, Chile, on 7 December. That crashed while on its way.[10] A second was immediately sent as relief but, as it emerged after the meeting, this too was damaged. A third plane eventually arrived safely but too late to be of any use.[11]

Hopes also faded that Rymill of the BGLE might help. Wordie, who had been the first to float the idea of enlisting the aid of the Graham Land expedition, began to query this possibility himself. He now felt that Rymill, with his ship the *Penola*, could not manage to search in the region of Charcot Island until much later in the season. Lying in the Bellingshausen Sea at almost 70° S and at 102° 20′ W, the ice at the island's edge would still be too tightly packed.

Hugh Millard in the American Embassy forwarded a copy of the response Rymill had made to the US. Arriving a few days after the emergency meeting, it confirmed what Wordie feared: though the ice in his area was clearing rapidly, there was still too much of it for *Penola* to move far through floes that were still heavy. If they had been situated further round the coast, they could have helped, but it was not possible to assist from where they were. *Penola* was not fitted to work in the pack ice she would encounter to the south of Graham Land. Perhaps, Rymill suggested, it might be better to use an aircraft working from Ellsworth's ship, *Wyatt Earp*, herself. Interestingly, he added that he was in daily communication with *Wyatt Earp* and had not appreciated that Wilkins needed any help. He had assumed, as subsequently proved to be the case, that lack of news from Ellsworth was due to a fault with his wireless.

Chapter 9 Many Masters

The third possible alternative also closed. On 7 December, the US Embassy forwarded a reply from Wilkins on board *Wyatt Earp,* in which Wilkins, who had at first rejected any offer of help from the Discovery Committee, suddenly changed tack and accepted gratefully. Wilkin's reply reached the Committee in time for their Emergency Meeting on 9 December. With not only a simple agreement but a firm call from the leader of the expedition himself now on the table, the Discovery Committee could hardly stand back any longer. They may well, however, have been surprised by the details of what Wilkins now wanted. In the telegram where Wilkins finally accepted help from the London Committee he requested that *Discovery II* to go as near to the Bay of Whales as ice permitted and to listen for possible messages from Ellsworth from there. By being as near as they could to Ellsworth's most likely position, the telegraphist could at least then listen for the airman's radio trail set, if this were active. Were that radio to fail, the telegraphist could try to tune into Ellsworth's other radio. If necessary, wrote Wilkins, *Discovery II* could then return to New Zealand and take on board an aeroplane there. *Wyatt Earp* would, all being well, arrive at the Bay of Whales between 22 and 25 January 1936.

Wilkins made it clear that, while he was anxious not to interfere with *Discovery II*'s scientific programme, he would be grateful if *Discovery II* could remain in the region until the middle of January. He stressed that the research ship would only be needed for this if *Wyatt Earp* were unable to get there in time. In this case, *Discovery II*'s men might be asked to help further. In addition, Wilkins added, it would be helpful if *Discovery II* and *Wyatt Earp* could be in direct radio contact with one another once *Wyatt Earp* was far enough west to be within calling distance. In the meantime, until such a link could be established, the only way to communicate would be three-way, between Australia, New Zealand and the research ship. They would resort to this only if there were news of Ellsworth. Wilkins added that the last message from Ellsworth's aeroplane had come from 75° S 79° W. Wilkins made it clear that he thought that the first week of January would be early enough for *Discovery II* to pick up an aeroplane. For this, New Zealand would be the recommended place to find the plane, should one be required.[1]

Wilkins' requirements were many. The Discovery Committee listened carefully to each one. However, as they pointed out, unless an aircraft could penetrate some distance inland, it was doubtful whether such a search would be of any use. Unless some messages were received, there seemed little chance of locating the missing men. Even so, everything possible ought to be done both at the Bay of Whales and in the area of Charcot Island on the western side of the Antarctic Peninsula. There should be a two-pronged approach to Ellsworth's rescue. On the one hand, *Discovery II* should be dispatched with the best available aeroplane and proceed to the Bay of Whales. At the same time, the UK Government should let its counterparts in the United States know that it was doubtful whether Mr Rymill could help in his area. The Committee suggested that, before setting out for the Bay of Whales, Wilkins should try to explore the area south of Graham Land himself.

The Discovery Committee seems to have been completely unaware of Ellsworth's own contingency plans. It was therefore ironic that Charcot was the first place Ellsworth had wanted Wilkins to set up a base for himself, and for where his pilot should aim were they to come down soon after taking off. In one sense it was a wise choice, as it lay on the line proposed by Ellsworth for his route along the Graham Land Coast to the Stefansson Strait and then onwards to the Bay of Whales. However, Wilkins seems to have rejected the plan that he should establish a base there, perhaps with good reason. He had flown over the area in December 1929, and surveyed it from the air. He recognised Charcot as an island, even though, at the time of year he was making his flight, it was connected to the mainland by the ice shelf. He realised that in the same month, December, now 1935, it would still be so linked. Clearly, he realised, it could be difficult to know any exact position where he might best establish a contingency base camp that would be viable, whether for Ellsworth and Kenyon or for his own team to reach, once ice had begun to melt. He may have been wise to ignore Ellsworth's instructions.

As they mulled over the suggestion of Charcot Island, John Edgell, the Vice Admiral, began to grow impatient and summed up the situation briskly. It was clear to him that there was no chance of the ship making communication with Ellsworth for the time being. The Committee should simply press on, make the necessary preparations without delay for transporting the proposed aircraft, ship the aeroplanes, then take them south on board *Discovery II* at the same time as continuing to listen for signals. Better to have the aeroplanes aboard from the start rather than find later that they were indeed needed.

Edgell was well aware that the line of ice would be different each year and that it would be impossible to predict it as accurately as they would have liked. He now focused on the descriptions Richard Byrd gave in his writings about his journeys in the late 1920s and early 1930s. This would continue to concern him, so much so that at another meeting, days later, he fell to doodling on the back of an envelope, sketching the different lines the ice had taken over the last few years and indicating the various lines the ice edge might now take. He had read accounts of Captain Scott's experience and that of other previous explorers, as well as of Byrd's own encounters with ice only a few years before. He knew well the historic crossings recommended for those who were going to work their way through the pack ice at the entrance to the Ross Sea. Supplies were another point that concerned Edgell. If anyone did reach Little America, what would they find there? The answer to this was unclear. He felt sure that some would be available at the Bay of Whales, but they should still double-check with Byrd. 'Ask Byrd', Wordie then scribbled in the margin of his papers. Continuing his reflections, Edgell estimated that it would take until around 28 December to make the alterations to *Discovery II*, ship two aeroplanes, and reach the Bay of Whales. It was time to grip the situation and set the preparations in motion.

By the end of the meeting, the Committee finally decided that *Discovery II* should put in at Melbourne, take aboard one or more aeroplanes, proceed to the Bay of Whales as quickly as possible and, when there, see if they could pick up messages from the two Americans they hoped to rescue.

Just two positive suggestions had emerged: on arrival at Melbourne, *Discovery II* should either have the necessary alterations made with a view to carrying an aeroplane at a later stage, or the ship should take aboard one or more aeroplanes and then proceed to the Bay of Whales. Whichever of these plans was adopted, the ship should proceed to the south as soon as possible. The Committee also knew by now that the New Zealand Government was prepared to share the cost. They were also comforted to know that the General Post Office would arrange through the Portishead Transmitting Station near Bristol to listen for messages from *Discovery II.* As the situation was urgent and it was not practicable to wait for a new meeting of the Committee to be called, the chairman should be given the authority to make any decision necessary. From then on, they would keep meticulous details of costs, making sure expenditure for the Ellsworth Relief expedition was kept well apart from normal expenditure.

It took the Committee eight days to expand what lay behind their first peremptory telegram to *Discovery II.* Now the Committee agreed that it was time to inform their research ship more fully of their decisions. Two days later, on 11 December, copies of the Minutes of the meeting of 9 December reached Hill, as did a telegram despatched on the same day.[2] The telegram brought the first of many detailed instructions and contained much advice.

The Committee, it read, wanted to cooperate fully with the Commonwealth Government of Australia and join in the search for Ellsworth and his companion. On arrival in Melbourne, Hill should consult with the Commonwealth Government about shipping aircraft and the required flying personnel. Then he should head for the Bay of Whales, as far as ice allowed, to listen for wireless communication from Ellsworth. If they had to make any major alteration to take the aircraft aboard, Hill should first report the outcome of this meeting to the Committee, then await their decision. Slipped into the first paragraph of the message, ran the sentence:

> In this connection standing directions concerning navigation among pack ice should be adhered to.

Hill should report back to the Committee the times and dates of procedures and give them some indication of the date he expected to leave Melbourne for the South. This would enable the Committee to send instructions as and when necessary. He should also report the names of the flying personnel who were to board the ship. Then followed a correction to the details of Ellsworth's wireless, its frequency, its call sign and the times the explorer was scheduled to transmit. Wilkins, on *Wyatt Earp*, would be trying to communicate with Ellsworth when in calling distance of *Discovery II*.

From then on, telegrams from the Discovery Committee flew in profusion to *Discovery II*. So far, it would seem, nothing had been left out. Hill was to send details of *Discovery II*'s position at noon each day. He was to report in full any requirements and to seek permission from them to make alterations. As the Treasury had agreed to reimburse the Discovery Committee for the work, Hill was to keep a careful check on this aspect of their expenditure and note in detail all money that was spent on the relief expedition.

… … …

It was now general knowledge that the governments of Australia and the UK were working together to provide a relief party in aid of Ellsworth and Kenyon. National newspaper reports continued to keep the world well informed. Every regional paper provided coverage of its own local 'hero' where someone from their local community was involved in the search for the two lost American flyers.

To those more intimately involved, it was also becoming increasingly clear that there were many leaders. The Australians called the Expedition theirs.[3] The Discovery Committee in London never questioned that it was they themselves who should have the final word. It would be for Hill to handle any conflict of commands that might issue from the two governments. It also lay with Hill to have the final word in resolving the question whether he could or should take *Discovery II* to the Ross Sea at all.

Chapter 10 *Discovery II*: Her Ice Experience

The need to make a diversion to find or rescue people who were known to be in difficulties or who were missing in the Antarctic was a given. Everyone did what they could to aid another ship or expedition in difficulties. But the destination for this latest diversion was at least strange and, at most, alarmingly hazardous.

A sentence, in the first paragraph of the Discovery Committee's telegram on 11 December, puzzled Hill immensely:

> In this connection standing directions concerning navigation among pack ice should be adhered to.

Everything else Hill had read in the telegram made sense, apart from this. If Ellsworth needed help, then of course they would have to make a diversion in order to help him. But, from the moment they knew they were eventually to make for the Bay of Whales, Hill saw serious difficulties in reaching their destination. He shared his concerns with Deacon and the other officers. The Bay of Whales, an ice-locked bay some 10 miles wide, was the world's most southerly harbour. It had been created as two ice systems in the vast ice shelf had advanced unevenly, surrounding a rise in the bedrock which in 1934 Richard Byrd named Roosevelt Island. The ice systems then converged at the island's most northerly point. Lying at 78° 50′ S and some 1,500 miles to the south of Dunedin in New Zealand, the Bay of Whales still teemed with the whales for which Ernest Shackleton had named it in 1904. It was well known as an important starting point for numerous expeditions. Amundsen, Scott and Shackleton had all used it as the base for their journeys into the interior. More recently, in 1929, Richard Byrd had established not far inland, at 78° 12′ S 162° 12′ W, the first of the settlements he used as headquarters for his own exploration.[1] To reach the Bay of Whales was an immensely attractive idea, but one normally quite beyond the hopes or expectations of either marine staff or scientists on *Discovery II*.

Now it was clear that the Discovery Committee and the Australians expected the Bay to offer a base in the search for Lincoln Ellsworth and Herbert Hollick-Kenyon. It would be by far the furthest south either of the Discovery Committee's ships, *Discovery II* and *William Scoresby*, had so far navigated. To reach even the Ross Sea meant a journey, often hazardous, of over 500 miles through what was usually very heavy pack ice. Once in the Ross Sea itself, sailors could expect calmer conditions, but the pack ice they would encounter on their way there was a constantly changing environment which had challenged, if not destroyed, many ships.[2]

The pack ice which lay at the entrance to the Ross Sea would have formed mainly in the previous year when, from around March to September, cold winds from the south swept over the water, causing the Ross Sea to cool rapidly and to form ice crystals. These crystals in turn would have massed to become pancake ice. As the new areas of ice collided, they created a still larger stretch of solid ice which would only be broken by wind and heavy sea. Snow from above and freezing water below soon led to large expanses of ice which could stretch for a distance of one hundred feet or more. Broken up by wind or swell, these areas of ice then re-formed as even larger ice sheets, or floes, which might well stretch for many miles. Alternatively, they might crash against each other to create huge ridges where the blocks had been upended on impact or piled one on top of another. When one ice block crashed against the side of a ship it could easily push a vessel onto another block. In no time a ship could be trapped between them.

Sometimes the floes were so hummocked that even the Scandinavian whalers, well used to pushing through the ice, were forced to withdraw. The newer factory vessels, though specially built to tackle the heaviest ice, often experienced battered plates and loosened rivets, and returned with tanks full of sea water rather than the whale oil they had intended to collect.

The timing of a passage through the pack was critical. The ice edge retreated or advanced according to the season, and conditions varied from year to year. In some years it had extended well to the north. Its southern edge, bordering on the Ross Sea, was equally variable.[3] Its extent often took sailors by surprise. The safest months for penetrating this pack ice lay between late January and February to early March. This time was also the best for finding ice-free water in the Ross Sea itself. Even this two-month period of clearer seas could prove to offer too short a short window of time. If a ship did not find a lead through the pack, she could be trapped in the ice for long periods. The journey could take anything from five days to five months or more. In January 1840–1841, Sir James Clark Ross, for whom the Sea was named, had only taken four days to move through the most challenging part of the pack on his journey south from Hobart to the open sea which would be named after him.[4] Other ships had been less fortunate. *Southern Cross*, which took Carsten Borchgrevink to Cape Adare in 1898, was stuck for six weeks in heavy pack ice to the west of the Ross Sea. In March 1912, *Terra Nova*, another well-found wooden vessel and the lead ship for Captain Robert Scott's 1910–13 expedition, was forced to withdraw because of intractable ice conditions. Only six weeks before, *Terra Nova* had taken Scott's Northern party of scientists southward from Cape Adare, along the coast of Victoria Land to Evans Cove. Then, unexpectedly heavy early ice made *Terra Nova's* return journey to Evans Cove impossible, and the scientists had to spend the winter in what proved to be appalling conditions.[5]

Another ship, *Aurora,* drifted towards the north-west locked in ice for 15 months before she was able to escape.[6] This was the ship which Ernest Shackleton had intended would meet the main party of his trans-Antarctic expedition of 1914–1917, once they had crossed Antarctica from the Weddell Sea area, to provide them with fresh supplies. While Shackleton's overland party on board *Endurance* was held fast in the Weddell Sea, some of the relief team on the other side of the continent were trapped on board their own ship, *Aurora*. The ice which surrounded *Aurora* then broke off from the main floe. Instead of being prepared to meet the main group intended to arrive overland from the Weddell Sea area, most of Shackleton's Southern party drifted northward for fifteen months, trapped in ice, and leaving others in their group marooned on the main floe. The fate of these parties from high profile expeditions became part of Heroic Age history. To navigators who followed in the same part of the Antarctic coast, it proved the difficulty of predicting the arrival and the movement of the pack ice.

Other problems faced those who tried to make their way through the ice. Once a ship was caught between two floes, the screwing effect of floe against floe could combine to pull the ship apart. The noise was stressful for everyone. Below decks each clash of the ship against ice sounded like a gunshot, each crunch like the tearing of metal. Added to this, and most importantly, *Discovery II* was made of steel. She was an excellent ship for her intended purpose. Her entire frame had some strengthening for work in ice, while the plates below the water line had been strengthened even further.[7] But she had neither the superior strength of the new whaling ships nor the resilience in ice of the older wooden ships, a point which arose more than once at Discovery Committee meetings.

In Melbourne, Hill would claim that *Discovery II*, though made of steel rather than wood, was well up to the task of negotiating the Ross Sea pack. The research ship used regularly to sail southward to the ice edge to collect data and material for their investigations. Sometimes she entered ice for protection from wind or swell, or to collect scientific data. This manoeuvre itself, however, involved some risk. Ice floes could mass at any time, and on some occasions the ship's captain drew back feeling prudence the safer option. It was customary for the ship to sustain some damage from close encounters with ice. The ship's rudder or propeller might be damaged, the scientific gear trapped or completely destroyed; the Chernikeef log, which trailed below the water level to measure speed and distance, could become tangled with ice. There were limits to the amount of ice-infested water masters of *Discovery II* were prepared to penetrate. It was surprising, to say the least, that the Discovery Committee should support an expedition into pack ice as notorious as the 500-mile barrier to the Ross Sea. To say that Hill should adhere to standing directions and navigate through this pack made no sense whatsoever.

....

As Hill discussed his concerns with the others, memories surfaced of previous unexpectedly demanding episodes in ice. The scientists, George Deacon and Frank Ommanney, remembered the first of these brushes with heavy ice as far north as Bouvet Island in the South Atlantic at 54° South in late 1930.

On that occasion *Discovery II* had left Cape Town in September 1930 and was making for South Georgia with the intention of doing some survey work on the way. At first, they were beset by storm-force winds from the south-east and had to remain hove to for several days. Writing in his journal, Leonard Thomas recorded that these winds were the worst they had experienced for years.[8]

After the storm subsided, they spent a few days doing useful survey work, taking sightings of islands and skerries in the area. They then set course for South Georgia.[9] The first signs of ice appeared in the morning of 15 October, with growlers, brash and snow surrounding them. Later in the day they recorded seeing several whales. Two days later the ship reached Bouvet Island, an isolated volcanic island some 1,600 miles from South Africa and a similar distance from South Georgia.[10]

They first met pack ice on 20 October, blown onto them by a wind from the south-west, the ice unusually solid for the latitude and for the time of year. For a week the men grappled with conditions they had never before experienced in those regions. Penguins coasted by on icebergs, whales appeared ahead of them, seals and sea leopards passed by, stretched out on the ice. It was clearly an abnormally cold spring.

Fog then became an issue, seriously restricting their vision. At one point they suddenly realised they were in the lee of a large, tabular iceberg blown almost onto them by the wind. Later, as visibility began to improve, they saw numerous bergs, both large and small, all round the area. These were all the more alarming as, with the screen on the bridge iced up and the wipers only partially effective, they only sighted them at the last minute. Some of the larger bergs were tightly lodged in the pack, along with massive chunks of ice which had broken away from another iceberg. Snow fell steadily and, even when the fog lifted, visibility was often poor. One day a small berg drifted under *Discovery II's* stern. The ship was pitching heavily at the time and struck it with force. It was almost impossible to go more than 100 yards without having to stop. On one occasion they remained stationary for as long as two hours.[11] The only comfort lay in knowing that there were two factory ships nearby, though it is questionable how much help these could have given in an extreme crisis.

When they emerged from the pack about 50 miles to the north of Grytviken in South Georgia, on 3 November, two days later than planned, it emerged that *Discovery II* had sustained a number of leaks, one in the oil fuel tank, others in the tank situated in the forepeak. Water had seeped into the bilges, leaving everyone on board almost speechless with relief as they realised their good fortune to have survived in the face of such damage. It had been a tense few days in a surprisingly late spring. As William Carey, captain at the time, put it. 'The ship behaved extremely well in what is probably the heaviest pack ice that one would permit her to encounter.'[12]

The ice remained heavy until the year end, but after some improvement they headed further south, believing that the ice would clear by February. They made for Peter I Island in the Bellingshausen Sea, hoping to identify land still unknown and perhaps chart it for the first time. However, they never reached the land they felt must exist somewhere to the south. To the west of Peter I Island, at 68° 51′S 90° 35′ W, they were soon confronted by vast quantities of pack ice. The captain persisted for some time, then *Discovery II* met the heaviest ice of all, floes up to 30 feet thick. At 69° S, and well away from the edge of the continental shelf, they were still at least 250 miles from the land they were trying to find. They remembered that Wilkins had flown south from that position for 170 miles, found the ice continuous and, admittedly in poor visibility, seen no sign of land. The lack on the horizon of the darkening of the sky, known as water sky, and the seemingly endless white ahead, made it clear that it would take an exceptionally ice-free year for any ship to reach the coast. This great stretch of impenetrable ice corresponded with reports they had often heard. It now seemed that, unlike elsewhere in the Antarctic, the pack ice in the Bellingshausen Sea did not normally move away from the coast in summer. Exploration to the west of Peter I Island, other than by seaplane, they concluded, could only be achieved in a year of exceptional ice conditions.[13]

The cruise to the Bellingshausen Sea provided a great surge of enthusiasm for future survey work, but it came at a price. The propeller frequently struck ice. Small leaks began to appear, letting water into the fore hold. The rudder hit ice with such force that it was thrust upwards by almost 90°. More than once they had to tie up to a large ice floe for repairs to enable the carpenter to caulk the leaking seams in the fore peak. Amazingly, the scientists still managed to complete a number of stations. There was great relief when they passed through the narrow entrance known as Neptune's Bellows and entered the large caldera which provided the harbour on Deception Island, made fast alongside a steamship, *SS Melville*, and were then able to pipe down all hands. Safe at last from the ice, a survey showed damage to both the plates and the propeller.[14] It was no surprise to find that this damage was serious. Everyone on board was well aware how fortunate they were that *Discovery II* had proved as resilient as she had.

Chapter 11 FOREVER ICE

To these accounts of *Discovery II*'s experience of untoward challenges in ice, Hill could add his own memories. He had joined *Discovery II* just a few weeks after the research ship returned from the Bellingshausen Sea. His first duty was as officer of the watch on a dull, overcast evening in March 1931 when they slipped buoy, left the smell of Grytviken's flensing platforms well behind, and proceeded to sea, first in a north-easterly direction. Hill had reached South Georgia on board RRS *William Scoresby* two days before and found his short stay fascinating as he met shore staff at Grytviken and saw the working of the new Discovery base at close quarters.

When he was off duty, he was soon regaled with accounts of the recent problems in Antarctic pack ice and listened with amazement and awe. Recently arrived in South Georgia, he had already seen enough of the snow-topped mountains and ice-filled landscape to marvel at its immense beauty, and he had felt a strong feeling of calm as he surveyed the scenery that was to become, for the foreseeable future, the main base to which he would return after long weeks at sea. He very quickly came to call South Georgia his southern home. While he remained optimistic, he now realised how unpredictable the ice could be.

He would experience this seeming capriciousness for himself less than one year later, when, once more in the Antarctic high summer, *Discovery II* experienced an even worse situation than she had faced in the Bellingshausen Sea.

They sailed southwards from South Georgia in early January 1932, on their way to the Weddell Sea. The chief scientist, David Dilwyn John, was hoping to find out why the water which flowed out of the Weddell Sea in a north-easterly direction was much richer in plant and animal life than the water to the west, which came from the Bellingshausen Sea. Each flow of water had very much the same temperature as the other and provided similar nutrients necessary for plant life. It would be helpful to find an explanation for a remarkable variation between the forms of life.[1]

They continued for six days, for a distance of 700 miles, taking a line of stations to the east of the South Sandwich Islands and steaming steadily southward through drift ice until they reached 70° S. Up to this point the ice posed no risk. The floes, though sometimes fairly large, were never heavy and left good leads between them. It was soon after this, as Hill recalled, early on 19 January, that their fortunes changed. Ahead of the ship, floes rose to 15 feet in height, creating a wall of impenetrable pack ice. This was topped by a surface of upended floes, pressure ridges and hummocks, all old ice which posed considerable risk. Hill remembered how relieved they had been when Carey called to turn *Discovery II* to the north-west, began to steam along the ice wall, and started to run a line of stations towards the South Orkneys.

But by midday that day, at 70° S 24° 20′ W, they were surrounded on all sides by hummocky ice, with floes firmly ridden one upon another. Round the ship, patches of thin ice began to form, soon to be consolidated with looser floes. Before long, continuous snow thickened the surface of the ice even further. Within a day, the sea had frozen over completely. The pack ice left hardly a lead in sight. Then followed a week when almost every moment threatened disaster, with *Discovery II* pinned in heavy ice, for much of the time virtually immobile. Heavy snow fell unremittingly so that they could barely see ahead. 'Butting' the ice, the traditional manoeuvre of reversing the ship and then moving forward at speed to force a path through the floe, could take up to two hours. These were tense moments for everyone, especially for those below decks where, as Leonard Thomas described in his journal, it sounded like a pistol shot every time the ship struck ice.[2] The shocks soon forced the rudder out of alignment. Next morning, 23 January, they felt another sickening jolt from the rudder. A list to port, on which they had already begun to comment, became more noticeable while, strangely, the rudder seemed undamaged.

On one occasion, as *Discovery II* moved slowly towards a crack between floes, a sharp, rending sound rang out. The ship suddenly heeled sharply to starboard, and fuel began to emerge from the ship, leaving a dark stain to spread ominously over the water and the surrounding ice. Other difficulties then beset them. When they secured alongside a floe with the small kedge anchor in order to readjust the rudder, one of the mooring wires fouled the damaged bilge keel. More leaks appeared, allowing sea water into the fore hold. Another time, when trapped yet again between impenetrable ice floes, they watched, as in slow motion, as the current drove an iceberg steadily towards them on their port side. It was one mile long and 120 feet high. A strong wind was blowing from their starboard side and pressed them ever closer to the berg. They witnessed the way the berg, pushed by the current and moving at a rate of about 1½ knots, created a strong wave ahead of it and caught *Discovery II* in the pressure of its arc.[3] Then, marvellously, the iceberg passed just astern of them, leaving the ship tossing dramatically in its large wake. Relief and wonderment combined in equal measure. As the scientist Dilwyn John commented after the event, this was a clear indication of the way icebergs, with their far greater depth, moved according to the current, whereas ice floes, largely on the water surface, would be driven by the wind. The powerlessness they all sensed as they watched the approach of the great berg emphasised all the more their vulnerability in the midst of so much ice.

Only after a full week of this hazardous progress, did the pressure drop and bring relief as a gale force wind began to blow from the south. The ice opened up considerably and on the 27 January, at 61° 30′ S 35° 20′ W, *Discovery II* was at last able to emerge into clearer water.

As they limped back to South Georgia, pushing slowly into winds which often neared gale force, they found more leaks between the plates. It still seemed touch-and-go whether they would make landfall safely. They rounded Cooper Island to the south-east of the island two days later and made their way to the north-west along the coast, pushing into winds of increasing force. Their relief was palpable as they tied up in Leith Harbour on 29 January, after more than a week in conditions that well surpassed those experienced near Bouvet Island and in the Bellingshausen Sea just over a year before.

On arrival at Leith Harbour, Captain Carey entered the usual Note of Protest, the Board of Trade requirement to outline the cause of any damage to the ship that might later emerge and to free the captain from responsibility for it. On this occasion, Carey simply wrote the phrase: 'ICE AND BOISTEROUS WEATHER'.[4] It was a phrase he used frequently, but on this occasion with considerably greater poignancy than usual. Using the upper case was the nearest reference he made at this point to the exceptional challenge of the journey back from the Weddell Sea, with ice that had all but crushed them, followed by a wind often near gale force, in a seriously damaged ship.[5] While the wind, now storm force, continued to batter the shore, the men saw to immediate, essential repairs to *Discovery II*, after which the Lloyd's Agent declared *Discovery II* sufficiently seaworthy to proceed, to Simonstown in South Africa. There more detailed checks were made, and full repairs carried out. It took even longer for the memory of their week in the ice of the Weddell Sea to fade from the minds of both sailors and scientists.

Hill did not doubt that *Discovery II* could withstand some difficult situations. The return from the Weddell Sea had not been the last challenge for the ship or her sailors. They had all experienced more than a few episodes since then. Working in streams of ice, forcing their way through unexpected floe ice, and even colliding with an iceberg, though fairly rare events, were all part of their experience. The weather could change dramatically and quickly. They might well quite suddenly find themselves driven by gales or storm force winds, battling heavy swell, or pushing through heavy ice, all in a short space of time.

Discovery II was, in the terms of Neil Mackintosh, the chief scientist of the third commission, a 'well found' ship and as robust as could be expected. Just over a year before, in 1934, they had experienced another difficult time as they were emerging from an area of pack ice. Mackintosh describes one incident with graphic detail when, after a sharp fall in the barometer, a gale rising to storm force blew from the west, and the swell drove them onto heavy ice. The lumps, Mackintosh wrote, were like London buses, pounding in the swell. As they pushed through them, *Discovery II* was tossed around, often hitting the ice with immense force.[6] She survived this battering with no real damage.

Hill's experience from this occasion, added to many others over the previous five and a half years he had served in *Discovery II*, had been invaluable. In later years he recalled that on many occasions he had experienced great fear. In spite of that, his strong faith had always given him confidence that they would survive. Instinctively, he had always remained optimistic. But he also knew that there were practical limitations to where and how they should proceed. Provided he was able to foresee the conditions ahead, he would certainly know when to draw back.

Now, in early December 1935, it seemed to him that the Discovery Committee's instruction to go through the Ross Sea pack ice, seemingly at any cost, ran counter to any good sense. They might possibly venture through it a month later when ice had melted further. At any time before then, Hill and his colleagues considered it almost impassable.[7] The three worst episodes sailors and scientists had experienced in *Discovery II* in ice-filled water, near Bouvet Island, in the Bellingshausen Sea and in the Weddell Sea, had all been unexpected. Their troubles when trapped in Weddell Sea pack had occurred when ice had formed in the spaces and in the leads between ice floes where normally they would have found a way through. That had been in high summer. Unexpected then, the same might so easily happen again. However much Hill might try to exercise his judgment at the start of a foray into ice, he might then be waylaid by unforeseen conditions. If he ventured too far into pack ice to make a safe exit, he could destroy in a moment what was in effect a 'tin can' of a ship. To enter the Ross Sea willingly before mid-January, when they already knew the risks, seemed utter folly.

It was not so much the fear of penetrating the Ross Sea pack. He remembered the delicate journey through pack ice to the east of the Ross Sea when they took relief to Admiral Byrd in 1934. They met various challenges on several occasions during the enterprise, but that journey had taken place in early February, when the Antarctic summer still lingered, and the ice was at its least for the year. The outcome of that expedition had been highly successful. The project now proposed might well prove the same, provided they could wait until the same month.

Rather, Hill's main concern lay in the conflicting messages which came from the Committee itself. This one issue, above all, now troubled him. The Committee was requiring him to take *Discovery II* through notoriously difficult pack ice to reach the Ross Sea and the Bay of Whales, before, as they themselves admitted, the season they considered the safest. Yet, on taking command, he had accepted that he should have due regard to important ice regulations which should effectively bar him from this very activity. Where he hoped for more support from the Discovery Committee, none came. The words of the Ice Regulations ran repeatedly though his mind:

> In this connection standing directions concerning navigation among pack ice should be adhered to.

The Ice Regulations, he had long accepted, were Standing Orders no responsible master of a ship of the Discovery Commission could or should ignore. Even if he, the one who would make the final assessment of conditions at the time, deemed it possible to make the passage through the Ross Sea pack, it would be a venture that contradicted all the received wisdom of the time. He expected far less ambivalence and much clearer support and understanding from the Discovery Committee than was apparent.

Chapter 12 Ice Regulations

The Discovery Committee first drew up their Ice Regulations and brought them into force in 1931. Until early that year, nothing had occurred to concern the Committee back in London about the strength and ability of their ships to work safely in ice. They had full trust in the masters of their ships and in their expedition leaders. It did not occur to them to question just how far into ice-infested water her master would take *Discovery II*. The new research ship had been built very much to fulfil the requirements and recommendations of Dr Stanley Kemp, the Director of Research. She had state-of the-art equipment and was faster than the first, wooden *Discovery*. There were still some who still hankered after the older ship but, iconic though the first *Discovery* was, recalling as she did the start of the Heroic Age of Antarctic exploration, she had neither the manoeuvrability nor the power of the new ship. Crucially, while being constructed with steel, *Discovery II* was strengthened for work in ice, though it was assumed that the ice she entered would not be heavy.

Early in 1931, two factors coincided by chance to lead the Discovery Committee to specify more closely how their ships were to view work in whatever ice she entered. They had recently asked the insurers if there were a way to reduce insurance premiums. The insurers' reply had been challenging. Even as news was coming through to London of the damages *Discovery II* had sustained in the Bellingshausen Sea in January 1931, a reply came from them, dated 10 February, which stated clearly that there could be no reduction in premiums. Already, the insurers claimed, the Committee's ships had incurred considerable loss, and the Committee's business had not, for them, been a paying proposition. Their premiums were as low as the Committee could expect, their terms as favourable as possible. Many underwriters might reject the Committee's business altogether, so specialised were the risks. Should *Discovery II* sustain less damage than the other two ships, they might reduce the premium on renewal.[1]

In April 1931, Sir Fortescue Flannery, marine engineer and influential member of the Committee, took up the case. The Discovery Commission's ships, he said, RRS *Discovery II* and RRS *Sir William Scoresby* could not be compared with the whaling and factory ships which were known to push deep into ice-covered waters. The Committee and their staff had plenty of contact with the whalers and the way they carried out Antarctic whaling. They knew well that the factory ships would often enter the ice, even if it meant they had to cover many miles of pack to reach the best whaling grounds such as the Ross Sea. Smaller whale catchers would accompany the factory ships, on occasions serving as ice breakers for the larger ship. All whale catchers, large or small, were at risk of being caught in the ice and damaged. The casualty lists of their losses showed this well.

Sir Fortescue added that, if the insurers imagined that the Committee's ships experienced similar risks, they might re-consider the matter. The Committee's ships, he told the insurers, were different. Though they were indeed well protected against ice, they never voluntarily entered it. The number of times they had done so was very few. Mostly, when out of harbour, they worked in open water with plenty of sea room. The risks they took could not be compared with those taken by commercial whalers. Their work was entirely different, not catching the whales but observing their habits and marking them in order to track their movements once they had been marked. For this the ships worked in open sea. The same was true when they took soundings and samples. The sum of all this, wrote Sir Fortescue, meant that the ships kept as far from the ice as possible, both for the reasons of safety and in order to carry out the proper work of the ship.

A second factor led the Committee to examine more closely the times their ships worked within fields of ice. Around this time, news began to reach them of *Discovery II*'s experiences near

Bouvet Island and in the Bellingshausen Sea. The risk to personnel, as well as the extra expense which arose from the incidents in the region, now struck them very clearly, and the Committee were forced to reconsider both the nature and the wisdom of the forays into what they called 'ice-infested' water. The Committee quickly pointed out that the Underwriters were assuming that the insurers were comparing the Discovery Committee's ships with those of the whale-catchers who went through miles of pack ice to reach the best whaling grounds.[2]

The Insurers accepted the point and suggested that they had gained their information from prospectuses issued when capital was being raised for the purchase of these ships. They pointed out that past claims had indeed risen from ice damage. Moreover, any steamer sailing in a region where she might be confronted by icebergs was at greater risks than one in open sea. The rate on steamers running to Montréal, for example, was always higher than that for those going to New York. Only the week before, *The Illustrated London News* had carried a photograph from the deck of *Discovery II* showing the ship to be, apparently, remarkably close to 'huge' icebergs, hardly a recommendation for reduced premiums. They would, however, present the case to the Underwriters when the policy was next due for renewal. Provided there were no claims in the next six months, they would do their best to get to get some reduction in the rate, especially if the new *Discovery II* proved more resilient than her predecessor, the earlier *Discovery*. The Insurers welcomed the idea of attending a Committee meeting, adding that they especially hoped that on that occasion they would be able meet the Master of the ship along with the rest of the Committee.[3]

Almost at the same time as this correspondence was passing between the two groups, discussions began about the damage *Discovery II* had sustained in the two recent episodes. It was now clear that, in the past year at least, their ships had certainly not remained outside the ice. Even so, by the summer of 1931 the vessel had experienced remarkably little ice damage, particularly when compared with whalers. *Discovery II* was the first of the Committee's vessels to sustain serious harm. The Ship Subcommittee established that claims would amount to about £2,000, with an additional £460 for scientific items.[4] While the Committee seemed not unduly concerned about the overall safety of the ship, they clearly realised how important it was going to be for them to have a sound ice policy they could present to the insurers.

Once Captain Carey returned at the end of the commission and could be present in person, the Committee began in earnest to discuss the issue. Their first such meeting, the 106th meeting of the Discovery Committee, was held on 2 July 1931.[5] The Committee chairman asked Carey whether they would be justified in saying to the Insurers that the Committee's ships, while very strongly protected, never voluntarily entered ice, and indeed had rarely done so. Carey contradicted their assumption, saying that *Discovery II* had, without doubt, voluntarily entered pack ice. Sometimes they needed to shelter from heavy weather outside the ice. On other occasions they went into it to seek scientific data.

Carey was confident that in another similar season, and especially with the experience he had gathered in recent years, he would not sustain a quarter of the damage. He now knew that, when the ship was in a swell, they must avoid hummock ice and loose pack. He would also avoid entering pack ice as late as March. But he stated firmly that there were occasions when it was safer to be within the pack than out of it. Others at the meeting realised that the captain should be flexible in his approach to ice navigation and accepted this. Their reasons were various. The chief scientist for the Discovery Commission, Stanley Kemp, felt it important to be able to enter the pack, as conditions within the ice floes were often of the greatest relevance to their research. James Mann Wordie, absent from this meeting, wrote that it would be unwise to hamper the captain by setting too tight a ruling: the final decision should be left to him alone.[6] At the next meeting, Wordie, who generally had an eye to a tactful response, confirmed this view when he was present in person. It was important, he stressed, that nothing hard and fast should be laid down in writing.[7]

The next meeting was inconclusive on the matter, but eventually, in September 1931, the Committee approved a resolution which satisfied everyone:

> The Discovery Committee are of the opinion that the question whether on any occasion the RRS *Discovery II* should enter pack ice is one which ought to be left to the discretion of the Commanding Officer.

The Committee stressed that the vessel should only enter pack ice for definite and adequate reasons. It might, they recognised, be necessary and justifiable to enter pack ice for one of several reasons. They might decide to enter the pack for purely scientific purposes; they might do so in order to avoid the long detours often required to circumnavigate tongues or streams of ice; they might need to seek shelter within the ice from an oncoming swell or strong wind:

> In all cases the Commanding Officer will have due regard to the apparent character of the ice he proposes to enter and will look primarily to the safety of his ship and crew.

The Ice Regulations were thus finally drawn up, to large extent in response to financial pressures, as part of the regulations for any seagoing master. Hill became fully aware of the Regulations from the moment they were published. When he took command of *Discovery II* in the autumn of 1935, just three months before the call to prepare for a voyage to the Ross Sea, he had been reminded of them and the clear warning they included. With his experience over the last five years, the Regulations made perfect sense. Hill saw no problem accepting this advice.

Now he shook his head in disbelief, barely crediting that, despite these instructions, the Discovery Committee was now sending him to the very kind of pack ice they had warned him to avoid. The least he expected was some modification to the regulations. He waited in hope, but no such telegram arrived.

Chapter 13 MELBOURNE

Hill and his colleagues shared their delight when, soon after 20.30 on 12 December, they sighted Cape Otway on the southern tip of South Australia, and cheered again when *Discovery II* made the Narrows at Port Phillip Heads by 10.00 the next morning. They reached Williamstown, the port for Melbourne, at 10.30 on Friday 13 December.[1] The midsummer warmth had been increasing for some days. In spite of their disappointment at having to abandon their research, scientists and mariners alike were relishing the freedom from their heavy, seafaring gear. But the welcome they received took everyone by complete surprise. While still in the Yarra River, the waterway leading up to the port, an aeroplane flew around them, taking photographs of the ship as she steamed up the river. Once in harbour reporters swarmed on the quayside nearby, even as she was still being tied up. Officers and men were already being received as heroes. This enthusiasm continued in the various press reports that continued to circulate all round Australia, some reports more accurate than others. Many exaggerated the enormity of the task that lay ahead of them. Reporters overestimated the time it would take to prepare the ship for the expedition. or they produced entirely their own version of events. 'Many of them', as an observer travelling with them in *Discovery II* wrote, though without giving examples, '[were] so amazing and ridiculous that it kept the Mess amused for days.'[2]

It was normal practice when in harbour for the officers to entertain the seamen to lunch and tea in the Mess. In Cape Town, on their way south just six weeks before, this had happened on most of the days. But now such courtesies were soon laid to one side and remained so for much of the time the ship spent at the quayside. Work had to start immediately and often continued well into the night. On occasions the men worked round the clock. Hill was grateful to have everyone's unstinted support.[3]

The scientists, George Deacon and James Marr, gave several broadcasts in which they described the ship's normal duties, and gave a general outline of the proposed search for Ellsworth and Kenyon.[4] Walker, the chief officer, organised the stowage of extra oil, provisions, water and all the other material necessary for the voyage, while Hill provided the liaison between the ship, the Commonwealth authorities and the Discovery Committee back in London. For this, Hill spent much of the time ashore in Melbourne away from *Discovery II*, meeting officials and visiting the various departments involved. Here, the Captain found frequent need for a degree of diplomacy and tact as well as attention to practical details as he met the various authorities to make the necessary arrangements.

Immediately on arrival, Hill was greeted by a representative from Macdonald Hamilton and Company, the Shipping Agents, and by a deputy for Captain John King Davis, the Commonwealth Government's Director of Navigation. Hill discussed the normal ship's business with the agents, deposited his Articles which guaranteed his certification at the Shipping Office, then reported to Captain Davis, who outlined in brief the requirements of the Australian Government. Hill and Davis then went on together for a meeting at the Defence Department. There they were formally received by two representatives of the Commonwealth Government, Archdale Parkhill, the Minister for Defence, soon to be appointed KCMG, and Malcolm Shepherd, Secretary for Defence. With them were two representatives from the RAAF, Wing Commander Cobby and with him Flight Lieutenant Eric Douglas, who had already been appointed the lead flying officer for the expedition. The programme for the search for Ellsworth was again outlined and details of the aerial search explained. Hill had caused some comment among his own sailors when he first appeared on deck in harbour dressed in his official attire.[5] Now, in the presence of this formal gathering, he was pleased he had the full support his uniform gave to his morale.

Several issues emerged in discussion, all of which required various degrees of determination, diplomacy, good sense, and wisdom. Foremost came questions of *Discovery II*'s suitability for the task and her ability to make her way through the Ross Sea pack ice. They discussed the amount of fuel they would need, and eventually reassessed the final route they would take. Hill soon found several considerations which, he realised, the Commonwealth authorities had not fully appreciated.

Firstly, he felt the Australians had not appreciated *Discovery II*'s limitations in pack ice, an issue which had concerned him from the moment the ship was diverted from the ice edge and instructed to head for Melbourne. The second telegram from the Discovery Committee, which gave expanded instructions for the enterprise, had reinforced that concern, with its reminder that he must keep the ice regulations well in mind. Hill now stressed to the Australians that the ship was designed for work in light ice only: in anything heavier, he was bound to make the safety of the ship and, above all, the safety of his men his overriding priority. If the ice conditions ahead looked too risky, he simply must not proceed.

Now, at his first meeting with the authorities in Melbourne, Hill's concerns about working a passage through the ice were very much in the forefront of his mind. For the time being, he did his best to explain to the Australians the full import of the difficulties they could encounter in pack ice. He made a quick sketch of the ship's construction to show her limitations, reading to them the Committee's Standing Ice Instructions and pressing his point hard. He would do what he could but, with *Discovery II*'s own construction and the Ice Regulations imposed upon him by the Committee, the constraints were clear. The Minister for Defence finally accepted Hill's standpoint as the basis for all future discussions.[6]

The question of making their way through the heavy ice which guarded the Ross Sea still left Hill with considerable concerns. He clearly remembered the meeting, barely three months earlier, when he had been taken through the forest of legislation and the various recommendations that encircled command of *Discovery II*. The Ice Regulations had been high on the list of points through which the Committee led him. Moreover, there was no mistaking that the second of the telegrams from London sent late on 11 December, ordering him to make for Melbourne with expanded instructions for the enterprise, included a clear reminder that he must adhere to them. What Hill, of course, did not know was that in a draft for the telegram of 11 December was a handwritten note in the margin, stressing that he should take no ice risks, arrowed to be inserted early in the missive.[7] James Mann Wordie had clearly emphasised the risks *Discovery II* would encounter if they attempted the passage through the pack too early in the season. There was undoubtedly a clash of interest: to go through around 500 miles of notoriously fickle pack ice at that time of year yet take no undue risks. In time, when drafting the telegram, the Committee saw the conflict in requirements and realised that, whatever the reason for *Discovery II*'s journey through the Ross Sea pack, they should stress the priority of safety for their men and their ship.

Hill knew that the only safe area would be the pack at the very entrance to Ross Sea where wind and currents exercised little pressure, but he had decided by now that he could at least start to work his way into the pack ice. He still hoped the Committee in London would alter the Ice Regulations for the project but made no mention of that at this point with the Australians. He chose for now to finalise that part of the conversation with the Australians. The call to continue or withdraw would be his and his alone.

Three days later, on 16 December, Hill sent a telegram to London, asking if the Committee would, after all, 'approve of careful attempt through the ice'. He pointed out that, while he felt he could do the job and navigate through the ice to reach the open sea to its south, there would be some inevitable damage to the ship. A telegram from the London Committee on the matter might, he hoped, bring the support for which he still hoped. The Committee's reply, which arrived after three more days, showed they would not offer it:

> … Committee unable to modify ice navigation directions communicated in my letter of 2nd October. In exercising your discretion you should bear in mind that ship is strengthened for work in light ice only.[8]

The telegram left the senior staff in *Discovery II* even more puzzled. Surely, they marvelled, the Discovery Committee had a better understanding of the nature of the Ross Sea pack ice and the way any change in wind or current could move it relentlessly, leaving a ship pincered between the floes possibly for days on end? It could hardly be called light ice. Why send them at all?

Ice was not the only concern to be discussed in Melbourne on 13 December. Hill also stressed the need for extra fuel, should the aerial search party have to make a forced landing and *Discovery II* be delayed at the Bay of Whales. He had to be sure that more oil could be shipped to *Discovery II* if required. While still at sea when making for Melbourne, he had already asked Davis for a whale factory to be on standby to re-fuel his ship at the ice edge when they arrived at the pack, and again when they left. There was nothing unexpected in his repeating the question now. The Commonwealth authorities assured him that they would despatch an oiler to him if the need became urgent.[9]

Over the next few days many questions flew backwards and forwards between Hill, the Melbourne officials and London. On several occasions, a difference of opinion would emerge between the Australians and the London Committee. For example, in one of the many telegrams to the Committee, Hill stated his need for a secure fuel supply. Borley, the Secretary in London, replied that he thought it would be highly unlikely any oiler would be in the vicinity. Eight days later, Borley wrote to the Southern Whaling and Sealing Co. asking if there might be any chance to re-fuel from a whale ship if there were one in the vicinity. Borley indicated that this was a purely informal request, and he certainly did not assume it would be granted.

The question of oil was crucial. A delay at the Bay of Whales was a strong possibility, and Hill knew he should carry as much extra oil fuel in drums as he could allow on board. Normally *Discovery II* carried the fuel below decks. Now it would mean using every spare inch of space on deck as well as in the hold. It would bring many inconveniences to the men's movement around the ship, hindering their steps and limiting the amount of technical equipment the scientists could take. It was, however, an inevitable decision, and could prove well worth their while if they were delayed at any point in their project.

The final total of fuel, for planes and ship, amounted to 120 tons. Around 80 tons of this was stowed on the main deck. The rest went in the fore-holds, while the large after-hold, normally reserved for scientific gear, was specially cleared to allow for even more.[10] Negotiations for this meant detailed telegrams, with queries and responses between Hill and the Committee, as well as careful discussions with the Lloyd's surveyors in Melbourne regarding the safety and stability of the ship. As usual, the Committee required precise details of the positions chosen for stowage. This amount of fuel was far more than they had ever carried before and, in the event, proved more than adequate. Even so, as a further precaution, Hill asked the Discovery Committee if he could change the route originally planned for them to take to the Bay of Whales.

Their original plan had been for *Discovery II* to go south following a great circle route and sail directly past Macquarie Island which lay some 1300 miles south of New Zealand. They would then turn south at 62° 30′ S 18° E. This, the shortest and most direct route to the Bay of Whales and the route recommended by Captain John King Davis, had been assumed from early on in discussion. It remained the assumption even as late as 20 December, when the Discovery Committee asked *Discovery II* to keep an eye out for Emerald Island, a small island, not yet fully charted, which lay on this great circle. Emerald Isle had first been reported in 1821, but not seen since. Its existence remained a mystery and, if at all possible, Hill should report its coordinates.

Back in London, the Committee relayed as much as possible to the press. For many readers around the country the tension and excitement were growing as proposed maps and positions were publicised. For the general public who were reading the almost daily update on the reasons, plans and proposals for *Discovery II*'s expedition, talk of a great circle may have been verging on the overtechnical. One newspaper tried to speak of a 'great circular' course. It was soon corrected, but not before it may have caused amusement to some.[11]

Dealing with the serious side of the matter, the Discovery Committee at least was unfazed by the term 'great circle'. They agreed that Hill should abandon that route and make straight for Dunedin in New Zealand, where they could finally top up with fuel before leaving for the south. It would allow a direct route south to the Bay of Whales.[12] Hill, for whom reassurance about the oil supply was crucial, was satisfied by this decision.

Gradually the various issues were resolved. In general, those present brought the practical discussions to a satisfactory conclusion. The first meeting at the Department of Defence had ended pleasantly, and an overall sense of good will prevailed. But difficulties of another nature began to surface, and for a while there seemed a strong possibility that the whole enterprise might grind to a halt.

Chapter 14 To Business

The crux of the problem lay in the attitude of the Director of Navigation towards Hill. A good relationship between the two was crucial, not only at that time in the planning stages. What they decided would lead also to the success of the entire expedition. While *Discovery II* was still at sea, Davis had sent Hill telegrams which, though demanding, showed the natural support one experienced seaman would offer to another. Now, in Melbourne, the atmosphere seemed to change.

Born in England but naturalised in Australia, Davis had been one of the most active of Australian sailors in Antarctic waters in the early part of the twentieth century. He had served as chief officer on *Nimrod* during Ernest Shackleton's Antarctic expedition of 1908–1909. Next, he served as second in command of Douglas Mawson's Australasian Antarctic Expedition in 1911–1914. In 1917 Davis captained *Aurora* after her return to New Zealand to go the rescue of Shackleton's Southern party, still stranded at Cape Evans on Ross Island.[1] *Aurora's* fate had already become part of the folklore which circulated as a warning about the dangers of ice navigation. For Davis, the success of his rescue operation provided a further tribute to his own skills and successes. More recently, between 1929 and 1930, Davis had been Captain of the original *Discovery* on behalf of the British Australian and New Zealand Antarctic Research Expedition (BANZARE). Three times a Polar Medallist, he possessed a detailed knowledge of Antarctic waters.

Problems were apparent from early on in discussions between Davis and Hill. Over the years Davis had acquired a formidable reputation for tough dealing, and word of his peremptory nature had spread. His manner of straight-talking and bluntness was known to many. Hill already knew something of this characteristic from what he had heard of Davis' negotiations with the Discovery Committee on behalf of Douglas Mawson when leasing the first *Discovery* for the BANZARE expedition. Hill was to some extent prepared for Davis' bluntness, but by no means to the degree he now encountered it. Perhaps Davis, used to the wooden structure of the old wooden *Discovery*, had been taken aback at the steel ship presented to him in *Discovery II*. Or perhaps he had expected to be meeting a captain much older, who by nature of his age exuded more apparent gravitas. This would have been ironic since Douglas Mawson, Davis' one-time leader and frequent sparring partner, once said that by the age of 30 a man was too old for Antarctic exploration.[2] Aged just 27, Hill's looks still suggested youth, and this too may not have appealed to Davis. Or again, Davis may have perceived him as a threat to his own ascendancy. Whatever the reason, Hill's earliest encounters in mid-December 1935 did not bode well.[3]

Three matters in particular led to explosive reactions from Davis. The first lay in the possibility of their having to make structural alterations to *Discovery II*. On Saturday, 14 December, Flight Lieutenant Douglas and Flying Officer Murdoch visited the ship to discuss how best they might stow on board two seaplanes, one the De Havilland Gipsy Moth, the other the Westland Wapiti. Shipping the Moth was a fairly easy task as this, the smaller of the two seaplanes, would fit onto the space normally taken by the motorboat on top of the sick bay. Shipping the Wapiti would prove more difficult. To begin with, they thought that this, the larger seaplane, could be stowed on the fore deck with no great difficulty. Naturally, Hill preferred not to make any major structural alterations to *Discovery II* to enable her to ship the Wapiti. This had in any case been an express request from the Discovery Committee when, the previous week on 10 December, they sent what were intended to be final instructions. Hill sent a telegram to London early on the afternoon of 16 December, saying he thought none would be required. To be doubly sure, he had scale models made of the Wapiti and the ship's fore deck and stood corrected when he found the forward stowage area in *Discovery II* that had been proposed, was, after all, impracticable.[4]

Initially, Hill suggested they remove the teak rail at the fore side of the wardroom deck. This would then allow the plane to fit safely into position, but there would still be little margin for safe handling. The discussions continued over the weekend, and Hill eventually agreed that, if the Wapiti were to be safely stowed, they would after all have to remove the after Samson post from the poop deck. Situated at the stern of the ship, the post carried a derrick which could lift as much as six tons. This was in regular use for lifting deep-sea nets and trawls, holding steady the collections sought by the scientists as they worked on their stations, making life much easier for those who brought them in and handled them.[5] The Samson post also came into use for loading goods onto the ship. Now, however, its position would have prevented the plane from being lifted aboard smoothly. Only once it had been removed could a dedicated rig for the Wapiti be set up. Removing the Samson post was, in itself, a moderately easy task. But Hill had to ask permission from the Discovery Committee.[6]

This was just another request whose answer would join the copious directions which flowed constantly from the Discovery Committee. The Committee back in London was answerable to the Colonial Office for all money spent, and Hill had strict instructions to check out all issues with the Committee, especially anything that involved extra expense. Questions of fuel, provisions, oiling, stowage and the route, all came under their scrutiny. It created a curious position in Melbourne for Hill, who had to handle with care the narrow course between courtesies and requirements. In particular, he was bound to do so before he took any action out of the ordinary. The removal of the Samson post was one such occasion. Simple or not, this would count as an alteration, and Hill's next move was to refer to the Discovery Committee for approval. There would inevitably be some delay. When Hill pointed this out to Davis, the older man swore volubly and protested in no uncertain terms: such an alteration was just a minor matter, he claimed, and could be done without reference to the Committee. Hill knew otherwise and telegraphed the London authority that day.[7] Approval arrived by return the next afternoon, but inevitably it had irked Davis to have to wait at all.[8] His irritation at the delay was understandable. His language, however, was clearly unwarranted.

Difficulties did not end there. Removal of the Samson post might have affected the ship's balance and entailed potential risk. Once he had informed the Commonwealth authorities that he had the Discovery Committee's approval, Hill arranged for the Lloyd's surveyor to work alongside him during the alterations. The surveyor supported Hill throughout the reasonably short operation, but his involvement in the decision provoked yet another outburst from Davis.

A third source of Davis' seemingly relentless fury lay in his conviction that *Discovery II* was already unstable, even before the loading of goods. Stability was crucial and, to ensure it, the Lloyd's representative remained present throughout the loading. It was he who had the final word on the ship's safety and, when the extra weight had been taken aboard, saw to all the measurements of curves and balances. To Hill's relief he gave full clearance. The support of a Lloyd's representative was essential at any time. Now it provided the key for Hill to disregard Davis' next and final objection, the ship's stability.

Davis realised that on this matter his bluff had been called. Nevertheless, he rounded on Hill yet again. This time he tried simply to deny that Hill could prove *Discovery II* was indeed truly stable. He would not mind in the least, he said, cursing volubly, if Hill's ship foundered. His only regret would come in having to lie to explain the loss. But by now Hill had had enough of the older man's unreasonable and obstructive objections. These last words of Davis proved too much for him. In Hill's own words, they could have seemed a knockout blow. He realised that his best move was to reply to what he would call Davis' broadsides in the same vein and resort to the same blunt language Davis himself was using.

Hill, in his turn, began to swear loudly. Anyone who overheard him at this point may well have been surprised at the otherwise seemingly equable young Scotsman's force and energy, also, perhaps, even, his range of vocabulary. Ironically, Hill's reaction proved the turning point in their relationship. By returning like for like, Hill at last won the older man's respect. From then on, Davis addressed Hill in friendly and supportive terms. Subsequently, Davis would give Hill important advice which the younger man, his one-time opponent, would value highly. Hill later described Davis as probably the most difficult man he had met. Once, however, he had established his own presence in terms that Davis for some reason respected, he could find him 'charming'. In spite of what Hill would call Davis' 'undoubted wickedness', he came also to hold him in high regard.[9]

Behind everything was the question where, ultimately, lay the final authority to whom Hill should refer. Davis' underlying irritation was understandable. At no point had it been established officially who was running the enterprise. Was it Hill, as Master of the ship? Was it himself, as representative of the Australian Government? Or was it indeed the Discovery Committee back in London, whose ship the Australians had requested and were now using? To find a third party involved, approachable only by telegram, was more than frustrating for Davis. But if *Discovery II* was to be drafted to attempt the rescue, it had, as Hill correctly knew, to be on the terms of the London Committee. It had taken *Discovery II*'s Master to hold his ground to make this clear.

Another task had come Hill's way which required a more delicate approach. Such was the interest and profile of *Discovery II*'s mission that several people expressed interest in joining. Early in the first discussions on 13 December, Hill heard that Sir Douglas Mawson, Antarctic explorer and most recently the leader of BANZARE, had offered his assistance in an advisory capacity.[10] The Discovery Committee had already been informed that he would be on board, in joint charge of flying operations, along with Flight Lieutenant Douglas. But, as Hill saw it, this was to be a straightforward expedition, well within the capacity of his ship and their personnel. *Discovery II* was already taking aboard more people than had ever sailed on her before. There would have been little room for such a distinguished person to swell the numbers even further. At the same time, Hill must have felt that, even though he would still have overall executive command of the ship, he would be constrained in making his own decisions. He therefore declined Sir Douglas' offer, saying that his ship's company was well able to deal with the work ahead of them. Mawson accepted the decision, and Hill sent him a telegram of thanks, relieved without doubt that the matter was so easily settled. Hill also turned down the proposal to take a newspaper representative.

Discovery II's popularity in Williamstown continued unabated. Onlookers swarmed to the quayside to gaze at the ship, and groups came aboard to be shown her layout, from laboratories to engine room. The personnel did their best to make them welcome. Officers, scientists and crew may have felt overwhelmed by their new-found reputation. In many ways, however, this worked in their favour. It certainly helped in one surprising way when normal ship's discipline had to be maintained. Soon after their arrival in Williamstown, Hill had to report one of the firemen who had been causing trouble. For the sake both of the fireman and of the other seamen, the man had to be kept under constraint. The police helped to keep the matter as quiet as possible, and the troublemaker was kept in gaol for a short time before being sent back to the UK.[11]

At last preparations seemed to be moving well, with good will on all sides. All Hill wanted now was some form of acknowledgment from the Discovery Committee of the risk there could well be for their ship if her Master tried to force a passage through the Ross Sea pack. The Committee had stressed to him the importance of their Ice Regulations. Now, Hill hoped, they might acknowledge the threat involved and the contravention of their ruling if he entered such notorious ice. They might, he felt, adapt the wording of the regulations accordingly, or at least give him cover in an official letter or telegram.

Chapter 15 A Full Stow

By the end of the next week practical preparations were well underway. At 11.00 on 18 December *Discovery II* moved to the dockyard pier. In the afternoon, dockyard workers joined *Discovery II* men to rig the platform on the after deck to carry the Wapiti. Work continued throughout the night and into the next morning.

On the afternoon of 20 December, they shipped the Moth, with its central section attached and the float undercarriage rigged. Once they had hoisted the seaplane aboard, they securely lashed it into position on top of the sick bay. Then they secured a new derrick just abaft of the main mast. Around midnight they completed the platform for the larger seaplane.[1]

A trial hoist of the 40-foot Wapiti began the next day. The plane was flown to the dockyard, then towed across the harbour to the quayside. However, a strong southerly wind thwarted attempts for the rest of the day. The following morning the wind abated, and, to the delight of onlookers, the plane was swung successfully into place, nose forward on the new stern rig. The airmen who had come aboard the day before now began to dismantle the plane's wings and to store them safely in crates in readiness for the sea voyage.

The two aircraft were destined for different tasks. The Gypsy Moth would normally have a tank holding 19 gallons. Now, with a second fuel tank holding a further 12 gallons, it could fly safely for four and a half hours. On this plane the pilot would not be able to communicate with *Discovery II*, so it would be used mainly for local reconnaissance and to provide the Captain with details of local conditions. If required, it could be used to locate the Wapiti or anyone who might be in difficulties within 130 miles from the ship. Normally it would never go out of sight of the ship. The Wapiti, however, would carry 165 gallons and be able to cruise at 100 mph for between six and seven hours. It would be well equipped with both shortwave and longwave transmitters, a directional gyro steering compass, sun compass, drift indicator and magnetic compass, and would carry much else in the way of skis, tools, cooker and stove. It would be able to go inland on a pre-arranged course of around 200 miles out and back, with a possible circuit of an extra 40 miles at the further point. Each of the two planes would carry parachutes. These would have attached to them containers holding enough food for two men for 24 days along with medical supplies which the American fliers could drop to the Americans if they located them.[2] Preparing the planes for safe transport during the voyage was another crucial part of the planning.

Then came the loading of the general food supplies. Given by the Marine Branch of the Department of Commerce in Melbourne, these were plentiful. If all went well and explorers and rescuers began their return to Melbourne in good time, what remained of the food supplies could continue as ordinary ship's stores until they arrived. What was left at the end of the voyage was to be handed back in Williamstown.

Now everything was packed in crates, each carefully marked to show its destination. No one could mistake their contents. A green band denoted that one crate was intended for the aerial party to use at shore base, should these be necessary; a yellow band carried by another was for the use of a sledging party. Some crates included provisions for the ship's personnel, others housed equipment ranging from a theodolite to assorted needles for repair work, from soft straw with which to stuff their boots for extra warmth to bars of chocolate and pots of honey. Well aware of the unforeseen circumstances which could so easily develop on the relief expedition, Hill ensured that they took on an ample supply for a land rescue party – three months' worth of tinned and three of fresh provisions. These were carefully chosen by the ship's doctor, Dr J.R. Strong, in accordance with current dietetic knowledge. The ordering and arranging of these supplies progressed smoothly and without hitch. Only later, after taking them aboard, did Hill realise that he had failed to pass on the details to the Discovery Committee. He hoped his explanation for carrying on board such ample provisions would still be accepted. They would, he explained, be enough to sustain four men for 60 days, and light enough in weight to be hauled by sledge. In an emergency they could be rationed and made to last a full season. Sir Douglas Mawson contributed a number of practical aids, including a sledge. These were to be returned to him after the expedition.

Early in the morning of 23 December, work began to secure the ship for sea. Fresh water tanks were filled, and the aeroplanes and gear secured. The central section of the Moth already rested on the roof of the sick bay, with undercarriage and floats resting in a cradle securely lashed into position beside it. The Wapiti cradle and fuselage was stored, fully rigged, with its nose facing forward on the platform erected over the after deck.[3] The motorboat rested on chocks erected in place of the starboard whaler, while the other whaler, normally stored on the port side, was lodged on the poop deck. *Discovery II* looked a strange sight with so much equipment on board. Aeroplane parts appeared everywhere, and an array of white-topped fuel drums covered any available space on deck.[4] Deacon ensured Hill had photographs to send to the Discovery Committee of the final stowage plan.

Early that evening, with their Lloyd's certificate on board and a copy on its way to the Discovery Committee in London, they were ready to go to sea. Even then there was more paperwork to be completed. Advice had been pouring in from every quarter as their departure drew nearer, and Hill had to respond to as much as he could. On 21 December Hill received from the Commonwealth authorities a copy of their instructions with an exhaustive list of their requirements. Among the 22 points itemised, Hill was requested to furnish a complete list of everyone on board, their rank and the names of their next of kin, and to make a check for stowaways. He should do everything possible to keep in wireless communication with the Director of Navigation in

Melbourne and should also make contact with Sir Hubert Wilkins as soon as possible. There were instructions for the airmen and prescriptions for the way in which publicity was to be handled. Other recommendations ranged from the approved speed to the date of their return. The list was exhaustive.

Hill was given complete authority over all others in *Discovery II.* He was responsible for everyone and everything. That apart, he could vary the instructions as he saw fit.[5] In reply, while still making it clear that his ultimate authority was the Discovery Committee, Hill agreed to do his best to comply with the instructions given to him by the Australian Government. He then sent a copy of them to London.[6]

Davis ended his own exchanges with Hill in a particularly pleasant manner. Gone was the time when he told Hill he did not care what happened to him. Now, Davis sent the younger man what proved to be invaluable advice on the best practice for managing journeys through the pack. He headed what he wrote as 'Notes re Passage of Ice Pack into Ross Sea', adding 'For the PERSONAL information of the Master. They are to be regarded in no sense as instructions.' Davis closed his missive with supportive words:

> I have noted these points as they occur to me because I believe you will read them in the spirit in which they are written. You have a difficult task and while I cannot do much to make it easier, I want to give you any aid I can. Wishing you all God Speed and a safe return, sgd. John K. Davis. [7]

Both the instructions from the Commonwealth Authorities and Davis' letter covering the 'Ice Notes' made it quite clear that Hill's judgment should be final in any decision whether to push forward through the ice or to avoid it.

Davis sent the Committee in London a copy of the advice he had given Hill. With it he also sent a covering note expressing, if not his irritation, at least his concern, at the conflict of control he saw between himself and the Discovery Committee.

He was, as he put it, unclear whether the voyage planned for *Discovery II* to go to the Ross Sea was to be conducted from Australia, or whether the Discovery Committee in London desired to maintain direct control over the vessel. He had therefore told the Master of the research ship that the instructions handed to him were to be regarded for the present as tentative only. He would always remain grateful for the assistance he had received from the Committee when he was sailing with the first *Discovery*, and so felt it incumbent on him to give the ship's Master some instructions before he left Melbourne. Once they had read them, the Discovery Committee could either confirm or modify his advice in any way they felt necessary.[8] Accompanying this letter to the Discovery Committee in London, Davis sent an outline of the procedure *Discovery II* should adopt. Here he once more expressed his concern that time might be wasted while the three governments lingered over the crucial decisions they were making. He saw no difficulty in sending the ship from Melbourne, provided firm pronouncements were made straightaway. If sailing were delayed while communications between the governments were still on-going, the Antarctic winter would be upon them before anything was settled. Lives, he implied, would be lost.

Referring to a tragedy of the mid-nineteenth century, Davis recalled that 40 expeditions had been sent out by various governments in attempts to rescue Sir John Franklin and the sailors in HMS *Erebus* and HMS *Terror* who set out to explore the North-West Passage in 1845:

> All were too late, and it is knowledge of what took place then which urges me to recommend that either a concrete plan should be decided upon at once, and the Master of *Discovery II* instructed to carry it out if possible or that the whole direction of affairs and responsibility for operations [be] left to the Discovery Committee in London.

Whether or not Davis still felt that the relief expedition should be controlled from Melbourne alone, rather than by the Discovery

Committee in London, is not known. Nor is it known how the letter was read by the London Committee in general. It was, however, a letter that was soon to prove useful to some Committee members in another way.

None of the underlying tensions in Davis' letter to the Discovery Committee were apparent to the rest of the world, and messages of good will continued to pour in for the captain of *Discovery II* and his companions. On the afternoon of the 24 December, Hill received the following message from the Minister of Defence of Australia:

> On behalf of the Commonwealth Government, I desire to express grateful thanks for the ready manner in which you responded to all demands made upon you and the expedition with which all arrangements for the journey to the Bay of Whales have been completed. The Government wishes you and the complement of *Discovery II* the season's greetings and a safe and comfortable journey on your errand of mercy which it is sincerely trusted will result in rescue of the American airmen unfortunately missing in the Antarctic.
>
> Archdale Parkhill, Minister for Defence.

Deacon and Hill replied in the same cordial vein:

> On behalf of the whole ship's company of *Discovery II*, I have the honour to convey to you our sincere thanks for your kind message. We are proud to be sent on this errand and will do our utmost to justify the confidence placed in us by the Commonwealth Government.
>
> Leonard Hill,
>
> Captain

They also sent greetings to the Discovery Committee on behalf of the entire ship. It was an amicable start to the Christmas season for *Discovery II*.

Chapter 16 Through the Bass Strait to Dunedin

Discovery II finally set sail from Williamstown in the early evening of 23 December. The ship's company mustered at the ship's boat station, the pilot came aboard, and they cast off half an hour later. The crowd of friends, relatives and well-wishers waving them off appeared to grow, even as they distanced themselves from the quayside and took the ship into Port Phillip Bay. Some two hours later, they slowed down four miles off Point Cook and rode to the port anchor overnight. Next day, knowing that their direction-finding apparatus would be more important than ever once Flight Lieutenant Douglas and his men started flying from *Discovery II*, Hill insisted that they spent time calibrating it, and in the mid-morning they took aboard a wireless expert from the Royal Australian Navy, Lieutenant Commander Newman, to help. They completed the tests within the morning, but in that time the wind had begun to freshen.

Soon after midday, *Discovery II* moved back to Point Gellibrand where Newman quickly disembarked. A fresh breeze was blowing from the south-south-west, and the pilot took the ship slowly through the Heads by the West Channel. He then moved them at full speed across Port Phillip Bay to Port Phillip Heads. The wind now blew from the north-north-east. By 16.20, when the pilot was ready to leave, it had freshened considerably. By this time a heavy sea was running, and the waves were already so much higher that Hill asked for a small amount of oil to be pumped overboard to settle the churning water. Only then could the pilot to get into his cutter and go ashore.[1]

Three hours later the wind, already a strong breeze, gave them their first real chance to test the stowage on board *Discovery II*. The additional weight from the extra goods they had taken on board caused the ship to roll even more than usual, but they had to secure just one storage case on the foredeck. This they did in the early evening at 19.40 on the first day. Apart from that, everything seemed firmly lashed into place, and from then on nothing shifted. By midnight, the ship was rolling heavily, and spray flew everywhere. The rolling would remain with them for much of the entire journey, so it was comforting to find how secure Lieutenant Walker's stowage was proving.[2]

The moment they reached the open water of the notorious Bass Strait, they met the full fury of a wind which moved constantly between south-west and south-east. Waves, as so often in those waters, were especially high. When they looked towards Cape Liptrap on the port bow, they noted several breakers which must have reached over 50 feet.[3] Even in the centre of the Strait, the waves caused the small research ship to pitch into them. Combined with the heavy rolling, this led to an unhappy Christmas Day for many people. Not for the first time Hill was glad to have his cabin on the bridge, with its greater access to fresh air. For the Australian airmen who spent most of the day in their bunks or leaning over the rail, it was an unfortunate start to the expedition. For the others, too, festivities were limited. But when not on duty, both officers and men managed to enjoy singing to the accompaniment of an accordion played by a Canadian sailor.[4] On a number of occasions they were forced to lie hove to for several hours. Inevitably, they covered little distance that day and shipped water the whole time.

The next day there was no change in the weather, and they experienced gale force winds most of the time. With their heavy cargo, aeroplane parts and cargo stored on deck, as well as in holds both fore and aft, any attempt at greater speed would have been unwise, if not impossible. Progress was again slow, and they covered less than four miles per hour on average. It was not until early afternoon, 27 December, that the wind began to ease. By midday on 29 December, they were still rolling very heavily but conditions were reasonable enough for them to make normal speed. The next day, with the engine running mainly at 104 rpm, for a while up to 105 rpm, they began to make up for lost time.[5] When Hill sent his daily cable to the Discovery Committee in London, he outlined the difficulties they had encountered so far but reported that they had sustained no damage and that spirits were running high. He added in positive note that, though it had proved a trying journey, no damage had been done, the wind was veering to the south, and *Discovery II* would reach Dunedin at 18.00 that day. This was duly reported in the press the next day for the loyal following in the United Kingdom. 'Two Days Behind Schedule' ran the heading in *The Times* on 31 December.[6]

A cheer went up on board ship when New Zealand's mountains eventually came into sight on the port bow. From the early evening of Monday, 30 December the duty officer began to mark off the lighthouses as they rounded the southern tip of South Island. First came Puysegur Point and its well-known rocks to the south-west and only 28 miles away, then Centre Island, less than 14 miles to their east. The lighthouses of Waipapa Point and Nuggett Point soon followed, the latter barely over four miles away.

The last part of the journey brought a particularly pleasant morning, when *Discovery II* moved steadily northwards up the coast of the south-east of South Island, New Zealand. Douglas for one, now completely fit, thoroughly enjoyed what he described as the lazy roll.[7] The ship rounded Cape Saunders soon after midday, and everyone relaxed as they saw the green, seemingly forested slopes of the waterway that led to Dunedin. In the early afternoon they stopped to allow a pilot to come aboard, then made their way up the estuary towards Dunedin. A medical officer gave them full clearance and freedom from quarantine at Port Chalmers, and the vessel was finally secured alongside at Dunedin by 17.00 on the last day of the year.[8]

Chapter 17 PENDING

One matter remained unclear even at the time *Discovery II* was leaving Melbourne. Hill was, knowingly, about to take 54 men in a steel ship into some of the most unforgiving and relentless ice conditions in the Antarctic. Yet even now he had not received full recognition from the Discovery Committee in London that the existing Ice Regulations could not be reconciled with such a journey. It seemed that, unless there was more to the silence than met the eye, the Committee was oblivious to the contradictions in their demands.

Hill had made the Australians fully aware of *Discovery II*'s limitations in ice but had insisted, in discussion with John King Davis, that he would nevertheless go ahead with the mission. Although there were enormous risks of wind direction and consequent ice movement, he was prepared to take on the challenge. Though by then he had Davis' backing for the enterprise, he knew well that, strictly speaking, he should follow Discovery Committee guidelines. He had sent the cable to London on 16 December, saying that he could hardly avoid entering the pack ice: he was reasonably confident that he could safely navigate through the pack at the entrance to the Ross Sea, but it was inevitable he would be risking some damage to the ship.[1] Even now, when he had already left Melbourne well behind, he still hoped to get the Committee's official approval.

When they had first received Hill's telegram, the Ship Subcommittee in London read into it two possible meanings: they could assume either that Hill was in genuine doubt about the regulations, or that he had had a dispute with Davis. If the latter was the case, Hill must presumably, they thought, be looking for official backing.[2] The Committee decided to assume the latter and sent their noncommittal reply as a result. They could not modify the ice instructions. It was down to Hill's discretion how far he should proceed, and he should bear in mind that the ship was strengthened for work in light ice only.[3]

The Discovery Committee was at all times severely constrained by financial considerations. This expedition to rescue the American airmen added a totally unexpected dimension to their expenses. The Committee secured some extra cover without incurring extra premium, but the Underwriters still required them to meet the first £5,000 and 15% of the excess. Hill put in a request for further cover for unloading goods and transporting them to Little America, should that be required. The Committee also agreed to grant further cover for the ship's personnel as well as for the ship's doctor should he have to travel by air in the case of an emergency. At a meeting on 17 December, they decided to ask the other two Governments to share the costs of the rescue mission that were not already covered. In response, the Commonwealth Government of Australia agreed to take responsibility for all the expenses connected with the aircraft and flying personnel from the time they embarked on *Discovery II.* To limit costs, it was agreed in general that none of the ship's personnel might take part in flights further than five miles from the ship, unless a serious emergency arose.[4] More than this the committee felt they could not do. For them to consider altering the Ice Regulations would damage what support they still had from the Underwriters.

It was purely their fear that they would ensue a permanent and undoubtedly unsustainable increase to their premiums which held them back from giving Hill the full support he deserved. But it was a blunt, if not callous, reply. If Hill had hoped for more positive support and the Committee had acknowledged that he could rely on their backing to enter ice that was generally known to be dangerous, he would have been satisfied. He naturally knew that he alone would have to make the final decision to brave the pack ice. That was an accepted part of his responsibility as Master but the Committee's repetition of the now familiar words had simply reinforced his sense of isolation.

Hill and Deacon were still puzzling over the reply and the time it had taken to arrive. The reasons for the London Committee's delay were not wholly unrelated. Members had other matters on their mind, as well as the dispatch of *Discovery II* to the Bay of Whales. Their reply would have been worded rapidly in the light of these other issues. Expense, as always for the Discovery Committee, was a major issue. The full Committee was looking for ways to reduce costs. Now the future of the first *Discovery* lay under their scrutiny.

The Ship Subcommittee had met on 11 December when they began to discuss the future of the original *Discovery*.[5] It was an old question, and as always would revive a variety of emotions. To many people RRS *Discovery* was still a potent symbol of what had become known as the Heroic Age, the epic days of Antarctic exploration. Since those times, even as recently as 1929, *Discovery* had been leased to the BANZARE Expedition, but now that that research was completed, she had once more been decommissioned. From then on, she had lain idle in St Katharine's Dock for over two years. Maintenance was proving expensive. It was a long-standing question how best to deal with her in her retirement from service. Just the year before, James Mann Wordie had written forcibly in favour of keeping her in readiness for some fresh work. If, for example, as was highly possible, the Commission's future research took them further into ice than normal, rather than have them remain at the ice edge, the wooden *Discovery* might yet come into service again. Far better, argued Wordie, that they should have the wooden ship for that, rather than the steel *Discovery II.* She might even, he suggested, be required at short notice for relief work.[6] This possibility was mooted now but soon dismissed. *Discovery* should be put up for sale.

The telegram sent from Hill in Williamstown reached the Subcommittee as its members were preparing to meet again, on 17 December. Their intention was to discuss insurance premiums, as always a further and heavy burden on finances. They were due to present their decision about the original *Discovery* at a full meeting of the Discovery Committee on the following day and were anxious to resolve that matter. Clearly concerned that if they altered the instructions for working in ice-infested water it would jeopardise their insurance policy, members quickly agreed that they could not relax the established regulations and, by so doing, seem knowingly to encourage work in heavy ice.[7]

This could well have been the moment for the Committee to see the paradox in their thinking. The old *Discovery*, they knew, was stronger and fitter than a steel ship for pushing a way safely through ice. Some Committee members remained of this view even after the meeting. They could, H. Horsburgh, the Technical Officer to the Discovery Committee, suggested, still prepare *Discovery* to go south as a standby should *Discovery II* become icebound. It might take at least a month to prepare her and would cost around £1,000, but the effort could be worthwhile in more than one way. For one, such a move might even make the old ship more saleable.[8] Whatever the motive, there was little doubt that the first *Discovery* was the more robust ship. However, when faced with Hill's new request for some relaxation in the Ice Regulations, the Committee's reply seemed almost a reflex action. Any change in the wording of the Ice Regulations could have a long term effect on the annual premium.

In the light of the background to the Regulations and the concessions they had already managed to secure, there was little else the Committee could say to any request for modification in the regulations. It was simply their fear that they would ensure a permanent and undoubtedly unsustainable increase in their premiums which held them back from giving Hill the full support he deserved and left him to take *Discovery II* through waters that held such recognisable risks without receiving, it might seem, complete support from them.

Their reply, offering no modification of directions, reached Hill in the early evening two days later. Whether at the time Hill fully appreciated the background to the refusal is not known. He undoubtedly felt a huge sense of responsibility in his position as final arbiter in the decision.

'Work in light ice only'!

It was a strange reply to come from the London Committee. In view of their lengthy discussions in Committee about the safety and the best time for the expedition to set out through the pack ice, members knew well the difficulties. Hill himself was above all concerned for the safety of his men and his ship. He remained puzzled and conflicted. As if the Ross Sea pack consisted of only light ice! He expressed his amazement to his immediate colleagues in no uncertain terms, before returning to the last stages of their preparations. By now, however, in spite of his misgivings, he had made up his mind. He realised that it was a waste of time to question the Discovery Committee further. While he remained deeply concerned, he would go ahead and take *Discovery II* south, through whatever ice he met.

At this point another element entered the mix. Before *Discovery II* was finally due to leave Melbourne, George Deacon, the chief scientist in *Discovery II*, had stepped in and approached the point from another angle. Waiting on his desk for him to complete lay an airmail to Stanley Kemp, now Director of Research on the Discovery Committee. On their last day in the harbour at Williamstown, Deacon added to his letter what he felt would be a discreet request; airmail, he thought, would be a route preferable to sending yet one more telegram by the normal channels. Not only would a telegram sent from *Discovery II* be seen by the telegraph officer and perhaps others on board, but also, crucially, by the Secretary in London, whose answer he realised only too well would be the same as before. An airmail direct to Kemp seemed much the best route. Deacon may or may not have told Hill what he was doing.

When Hill was first instructed to go to Melbourne, Deacon had instantly telegrammed his own compliance, adding that he would relinquish all stations immediately, but it remained seriously upsetting for him and the other scientists. Deacon asked for permission to do a selection of stations on the way. The first refusal would be confirmed in more detail in a telegram which Hill would receive when they were well on their way to Dunedin.[9] All research remained forbidden, a decree from London which filled the scientists on board ship with as much dismay as it had many of the Committee in London. Not only did the team on board ship face loss of time for their work but, as the scientists on the Discovery Committee in London also knew only too well, they would forego the sequence of their findings. Moreover, with the ever-present doubt of future funding for their project, they might never be able to complete the carefully planned programme of the current commission. Even if the Discovery Committee were granted time and money for a further commission, the cost of setting up the repeat programme would be great. Now, with Hill's doubts about the wisdom of the journey through heavy ice, and with no formal acknowledgment of its risks, Deacon saw a fresh cause for embarrassment were they not to go. After all the preparations, he wrote, it would be doubly disappointing should their overall leader, the one who held executive command, feel unable to take on the journey to rescue Lincoln Ellsworth and Hollick-Kenyon without, Deacon implied, further support from the Discovery Committee.

Hill was 'rather disappointed', wrote Deacon, by the Committee's lack of further advice or support: they were being sent on a route which would take them directly through pack ice, yet were expected to follow established Ice Regulations and put neither ship nor personnel at risk. Though Deacon had put it to Hill that a certain amount of experiment would be necessary, the Committee's last telegram had almost convinced Hill that he should not venture into ice at all.

In his airmail to Kemp, Deacon added that he was sure Hill would never enter the sort of ice that Carey had encountered in 1930 between Bouvet Island and South Georgia, or in the Bellingshausen Sea in 1931. With his experience, Hill would know well when to draw back and should be given the backing at least to assess the ice conditions. If they did not make at least some attempt, they would certainly achieve nothing. Hill had become so worried, Deacon continued, that he might simply refuse to go. Meantime, the RAAF had made considerable preparations. The Melbourne

Harbour Trust and Captain John King Davis had been helpful. Deacon highlighted the considerable publicity they had received throughout Australia. Reputations there would undoubtedly be lost should they fail to go. He might have added to his list the attention they had already received in Great Britain, where there had been almost daily reports in the press since early December. With so much publicity and, above all, their preparations and the support offered to them in Melbourne, their reputation would suffer to a serious extent with the public throughout the world if they failed to go. Added to all that, not only would they lose face, but it would be even more frustrating when they had already been forced to give up their scientific programme also to abandon the expedition. It would surely do the Committee a lot of harm if *Discovery II* did not make a serious effort to reach the Bay of Whales.[10]

What Deacon had seen and experienced in the neighbourhood of Bouvet Island and in the Bellingshausen Sea had been challenging enough. What he may not fully have envisaged was the 'sort of ice' Hill had met as a young officer in the Weddell Sea just the year after he joined the Committee, during which time Deacon was based in London. There were others on board *Discovery II* in December 1936 who had experienced those alarming days in January 1932. Deacon had undoubtedly heard of the situation they had endured, but he may not have appreciated just how much at risk the ship had been placed. Whatever he knew, his letter to Kemp was a powerful intervention.

The move proved worthwhile. Deacon's letter undoubtedly succeeded in alarming Stanley Kemp. But, with the intervention of Christmas and New Year, it took several weeks for the Discovery Committee to reply. It would not be until well into January that a final response reached *Discovery II*.

Kemp was first able to raise the matter at a meeting of the Ship Subcommittee on 7 January 1936. His news caused sufficient concern for another meeting to be called, and three days later, 10 January, they met again to discuss telegrams from the commanding officer of their research ship.

Kemp spoke forcibly, and Borley immediately saw the damage that could be done to the Commission's reputation if *Discovery II* did not try to go south and attempt to reach the Bay of Whales. This could well have been another moment for the Committee to realise the paradox in their thinking, or, if not in their thinking, then undoubtedly in the advice they had or had not given to Hill in *Discovery II*. The Discovery Committee, however, was always conscious that the responsibility for the expenditure of public money was theirs and theirs alone.[11] They felt they could not, even now, risk increasing the insurance premium which the Underwriters would undoubtedly demand.

Thoughts of the old *Discovery* being re-commissioned receded into the background, and members worked hard to justify their sending *Discovery II* to the Bay of Whales while not relaxing their stance on the Ice Regulations.

In the meantime, though Christmas and New Year had been free of official meetings, the lines between Borley in London and Wordie who was on holiday in Glasgow remained open. In another of his many letters to Wordie, containing the latest update on Committee affairs, Borley commented on Davis' correspondence which had recently reached the Committee from Melbourne, and which he had instantly forwarded to Wordie. Borley reported that members found Davis' notes about the ice passage most helpful. He referred to the telegram from the Master of *Discovery II* in which he had said that, though there might be slight damage to the hull, he could at least attempt a way through the pack. It was in the light of this that Hill had asked if the instructions regarding ice navigation could be modified.

At the time the committee had assumed that Hill wanted official backing from the Committee in his discussion with Davis. Now, however, wrote Borley in the light of Deacon's airmail, it appeared that Hill seemed very uncertain of his approach to ice navigation. Borley made no reference to any amendment to the Ice Regulations they might make. Instead, he wrote, Hill seemed to be taking his instructions in 'a rather exaggerated way' and should be 'heartened up'. With its suggestion of weakness on the part of the ship's Master, this seems a strange, indeed a harsh, comment from one who had been present at all the meetings where there had been immense focus on the challenges of the Ross Sea pack and the risks it could present.

A way through Hill's perceived hesitancy, Borley continued, lay in the notes on ice navigation which the Committee had read from John King Davis and which Wordie would have received in an earlier posting. Here Davis had declared that *Discovery II*, though a ship of steel, would be reasonably safe when the ice eased during the summer months. The wording of the veteran Australian sailor should give Hill the support he wanted without the Committee having to modify their Ice Regulations. Aware that Hill could already be at the edge of the pack, they were now anxious to send him a telegram as soon as possible. In this they could say simply that the Committee regarded Davis' notes as most helpful and in full accord with their standing Ice Instructions. With Wordie's agreement, the telegram could go straight away. Quite apart from any delay in this reaching Hill, the matter would be far better settled, Borley concluded, before the next large meeting on 10 January.[12]

Borley had already prepared a telegram to send to Hill before the main meeting on 10th. Wordie, who not long before had doubted the wisdom of sending *Discovery II* into heavy ice, must have agreed. With just a few alterations, the telegram was sent to Hill on 11 January.

By then *Discovery II* was already a week into her voyage, well on her way to the pack ice, and facing unknown challenges as she headed for the Ross Sea.[13]

Chapter 18 BRIEF RESPITE

Discovery II's stay in Dunedin offered the most relaxing time its crew had enjoyed since they were last in Cape Town at the start of the commission. On the evening of their arrival, New Year's Eve, a number of the ship's personnel, scientists, marine staff, and airmen went ashore and repaired to 'The Castle' on the slopes looking down on the city. There, along with many of the large Scottish population who lived in the area at the time, they enjoyed the spirit of Hogmanay to the full. Meanwhile, Hill remained on board ship to clear his desk before their final departure for the South. He welcomed the opportunity afforded by the relative inactivity elsewhere on the ship. For him this was a calm moment when he could catch up on the ship's business and deal with his personal correspondence. He started by checking accounts and answering some of the many queries the Discovery Committee had sent him asking for final details about provisions and other matters.

Then he turned to other correspondence. He picked up a telegram from Edward Nattriss, an active member of the Discovery Committee. As Shipping Officer, Nattriss had always taken a strong interest in the men on board the two ships, *William Scoresby* and *Discovery II*, and supported their interests where necessary. He and his wife had entertained Hill to tea the previous summer at their home on the outskirts of London, and Hill had pleasant memories of several meetings and discussions with him.[1] Nattriss had sent his good wishes to Hill and everyone on board ship for the voyage to come. Hill found it refreshing to have some personal correspondence rather than official material at this time, and he re-read the telegram with a sense of relief. To start with, he dealt with the practical details he had been asked to explain. Then he summarised the time in Melbourne, referring first to the flurry of initial questions from the Australian Government, and the time it had taken to deal with numerous other queries from people ranging from experts to the press. He had been, he wrote, 'inundated with experts and newspaper men' but felt that he and his people had left 'in fairly good grace'. He hinted at some sort of challenge from John King Davis, but only described this encounter in more detail in another letter to Edward Nattriss a few months later.[2] Then Hill touched on the more general state of affairs in *Discovery II*. He referred to the ship herself briefly, 'at the moment a funny looking ship with aeroplane parts everywhere'.

He went on to mention the officers on board with him, speaking highly of the Chief Engineer, W.A. Horton, and Horton's assistant, R.G. Gourlay. His two junior officers, Harry Kirkwood and Victor Marchesi, he wrote, were also excellent as right-hand men. Of another, alluding without doubt to the First Officer Richard Walker, Hill was more cryptic. He was deeply grateful for the way Walker had seen to it that the unusual quantity of extra stores had been stowed so well and could survive the fierce gales of the Bass Strait.[3] Without such competent and firm stowage, *Discovery II* might well have fared very badly. Walker remained an invaluable second-in-command, but the much more reticent and undoubtedly quieter Hill no doubt found his ebullient and extrovert manner exhausting.

Hill added his praise for the Australian airmen now on board, welcoming the good sense and the calmer, more realistic outlook that they brought to the enterprise. It helped, he wrote, that they toned down the excessive exuberance the regulars on board ship were now showing. As the first-in-command, Hill was deeply conscious of his duty of care for the airmen. He knew he would carry ultimate responsibility for them when the time came for them to take to the air above Antarctic waters and to fly over the continental mainland. The letter to Nattriss ended with Hill's description of his men's enthusiasm for the task ahead of them as, with a touch of dry humour, he recorded the latest quip to do the rounds on board *Discovery II*, that they were ready to 'sail to the South Pole if required'.[4] Clearly he had set his own qualms about disregarding the Discovery Committee's Ice Regulations firmly in the background.

The next day diverged from the original plan. Hill had originally intended to leave Dunedin immediately after they had shipped oil. However, on the voyage from Melbourne he had begun to have further doubts about the wireless direction-finding Apparatus. He still sensed this was inadequate for him to keep in touch with the Wapiti seaplane should a search flight extend more than 200 miles. As they approached Dunedin, he had telegraphed the New Zealand Shipping Company based there and asked them to send someone who would give advice on their direction-finder.[5] An advisor from the Amalgamated Wireless Association soon came and promptly recommended that the existing device be modified for use with aircraft. It was, as Hill had suspected, only suitable for the purpose for which it was originally intended, namely use at sea, and too weak for use with aircraft when operating from a distance. Simultaneously relieved to see his doubts justified, yet disappointed to have to postpone their departure, Hill agreed to spend another day in Dunedin while the machine was fixed. At least this extra day in port allowed everyone on board a last day to arrange their cabins and write their letters home.[6]

1 Officers, Chief Scientist and Doctor of RRS *Discovery II*, about to leave St Katharine's Dock, October 1935.

2 Chief Scientist of RRS *Discovery II*, George Deacon, and Master, Lieutenant Leonard Hill, November 1935, Simonstown.

3 Shipping the Gypsy Moth seaplane, Williamstown, 20 December 1935.

4 Hoisting the Wapiti seaplane on board RRS *Discovery II*, Williamstown 22 December 1935.

5 RRS *Discovery II* ready to leave Williamstown, 23 December 1935.

6 Seamen pumping oil to storage from deck.

7 The surgeon packing sledging rations into a parachute container, 5 January *1936*.

8 Flight Lieutenant Douglas and Sergeant Cottee repairing a sledge.

9 RRS *Discovery II* in Ross Sea pack ice.

10 A quiet evening in the pack.

11 RRS Discovery II alongside an ice floe, Ross Sea pack ice, 10 January 1936.

12 Nudging the ice.

13 Impasse.

14 Poling party at work, 11 January 1936.

15 Entering an open lead.

16 Approaching a patch of open water.

17 Flight Lieutenant Douglas (right) and Flying Officer Murdoch in Gypsy Moth, 11 January 1936.

18 Lowering the Gypsy Moth seaplane to a pool of water in Ross Sea pack ice.

19 Gypsy Moth taking off for reconnaissance flight, 11 January 1936.

20 Lincoln Ellsworth (centre) with party from *Discovery II* at Little America, 16 January 1936.

21 Entering the hut at Little America, 17 January 1936.

22 *Wyatt Earp*, sighted near the Bay of Whales, 19 January 1936.

23 Lincoln Ellsworth (left) and Hollick Kenyon on board RRS *Discovery II*.

24 Lincoln Ellsworth with Leonard Hill and members of the RAAF.

Chapter 19 SOUTHERN OCEAN

Discovery II left Dunedin on Wednesday, 2 January 1936, her store of oil fully replenished by the addition of 82 tons of oil, her personnel relaxed and eager to move on.[1] Hill welcomed the pilot on board at a little before 05.00, and they cast off just before 06.00. Hill's telegram to the Secretary in London, subsequently circulated to Edward Nattriss and Stanley Kemp as well as to H. Horsburgh, representative for the Crown Agents on the Discovery Committee, was as succinct as ever. He told the Committee that he had received a cable from Sir Hubert Wilkins, on board *Wyatt Earp*, who hoped to reach 70° S 170° E by 19 January.[2] In a normal year for ice, this point would be around the southern limit of the pack ice and could mean that the two ships might meet in the Ross Sea somewhere between the pack and the Bay of Whales. The main question in Hill's mind was the sort and the degree of ice he and his men might encounter in *Discovery II* once they had reached it. Before that, they had the Southern Ocean to cross. Hill knew that could present its own challenges.

They pulled away from harbour, and were soon sailing comfortably down the channel which wound through Otago Harbour. They left the green slopes surrounding Dunedin behind them, then passed Harington Point and Taiaroa Head with its sandy inlets to starboard. The sandy spit of the Aramoana mole slid away to port. Not long after they spotted sea lions, enjoyed the sight of albatross as they wheeled around the ship, and took in the expanse of open sea ahead. Before long, *Discovery II* turned southward. The familiar world of the Southern Ocean enveloped them once more.

Immediately, they picked up a westerly wind and returned once more to *Discovery II*'s inevitable rolling, a feature by now so well known to them all. They were in the area known as the 'Roaring Forties', generally considered the stormiest seas in the world.

They could have expected wild weather ahead, with westerlies as fierce as they had known at the start of their commission in the Indian Ocean and, more recently, in the Bass Strait. For the moment, however, they experienced no more than a fairly strong breeze. The rolling and pitching was moderate but acceptable, and everyone settled to their various tasks with easy familiarity. The Australian airmen, now well initiated into life at sea after their grim introduction at the end of December, accepted the ship's movement as normal. Even when the wind and sea later increased, the ship became drenched with spray and the rolling became heavy, those on board took it all in good part. By early evening the ship had reached the southern tip of New Zealand, and left Stewart Island well to starboard. Not long before midnight, Hill turned slightly to the south-west and set a course at 157° towards the Ross Sea.

While the weather was reasonable, they all got on with their work in whatever space they could find on deck. Some overhauled and rebound sledges, others made harnesses to be used with them. Others weighed out rations and packed them in the recommended portions. Steadily everyone took their own part in re-organising the space on deck, paying special attention to the large oil drums loaded there. On average, *Discovery II* used about 30 drums of oil a day, and what was stowed in the containers on deck was gradually pumped below. Seen from the bridge, the drums presented an unusual patchwork of different colours in the otherwise drab, grey sea around them. It seemed unfortunate to rearrange such a multi-coloured display, but both space on deck and the balance of weight on board ship were crucial.

The airmen did what they felt possible, at that stage of the voyage, to prepare their two aircraft. Soon, the old hands on board *Discovery II* got to know the Australian airmen well and to understand their different roles. Over the next six weeks good friendships developed between them. Some of the airmen were there for the maintenance and upkeep of the two aeroplanes. Only three of the seven were scheduled to fly the aircraft: Flight Lieutenant Douglas himself, Flying Officer Murdoch and Sergeant

Spooner. Hill and Douglas together worked on arrangements for the airmen's safety and made plans for relief and backup. Two days later the wind began to moderate, and they were able gradually to increase speed to eight miles an hour.

By Sunday afternoon, 5 January, the engine was working at full power and, despite the endless rolling, they kept up their speed of nine miles per hour both that day and the next. Around 20.00 on Monday, 6 January they crossed the 60th parallel and that night experienced darkness for barely two hours. This was the last time they had any prolonged darkness until they were returning to Australia almost a month later.

The next day, 7 January, brought mixed conditions. At first the rolling decreased, and they increased speed for a short while. Around midday they turned due south and, with the wind freshening, they slowed to a steady 8.06 miles an hour. The ship began to roll heavily, the effect becoming more pronounced by mid-afternoon. They now met increasingly heavy snow showers or sleet, and this, mixed as it was with the ocean spray, stung their faces fiercely unless they kept themselves well muffled.

Gradually significant pointers indicated that they were well south. Both the water temperature and the pressure dropped throughout the day. This was the area of the Antarctic Convergence, the physical boundary easily detected by such changes. By midnight, as the ship rolled incessantly and spray drenched them wherever they moved on deck, they realised that their steady push southwards had already brought them well into Antarctic waters. They could now expect far colder and more rigorous conditions. At least they were leaving the area of the Roaring Forties. Those who had made the journey to the Antarctic on previous occasions were used to frequent storms as they pushed across the Southern Ocean. This time, for once, the first few days had proved surprisingly steady and straightforward. Now they were pleased to be approaching the last stages of the greatest risk of gale force winds.

Chapter 20 Towards the Pack Ice

Spirits on board ship remained high. Nobody had yet mentioned aloud the word 'race' but by now a sense that they were in a competition had worked its way into the minds of many of them. Naturally, they hoped that *Discovery II* would reach the Bay of Whales before Sir Hubert Wilkins in his own ship, *Wyatt Earp*. In part it was a case of steel versus wood, the supposedly weaker ship proving its worth against the traditionally accepted stronger vessel. In part it was the natural desire to make their expedition worthwhile. Whatever the reason, there was a 'frisson' in the air. More than once in the diary he wrote on board ship, the Australian airman, Flight Lieutenant Eric Douglas, hinted at the idea of a race, subtly at first, then more explicitly.[1] Later, what had been a mere notion of a race became a recognised description of their push to the south.[2] The first suggestion of a friendly competition emerged around the time Hill had received Wilkins' telegram on 2 January, which told him that *Wyatt Earp* was expecting to reach 170° W 70° S by 19 January.

Discovery II finally established full wireless communication with Wilkins at midday on Sunday, 5th.[3] By this time *Wyatt Earp* was working her way along the west coast of the Antarctic, skirting the ice at 67° S 40° 122.23′ W, off the coast of Marie Byrd Land. *Discovery II* was then at 55° 11.9′ S 175° 5.3′ W, about 650 miles to the south-east of Macquarie Island. Each ship had almost 2,000 miles and unknown stretches of ice between themselves and their goal, the Bay of Whales. Hill replied, saying he would send their own position at noon each day and asked Wilkins to do the same: all their communications would remain personal unless the Discovery Committee asked specifically for details of Wilkins' coordinates. From then on, the two ships kept each other informed daily about their position. News of every update on *Wyatt Earp's* position sped through *Discovery II* as her personnel eagerly passed on details of their progress.

Soon too they could mark small stages in their own journey. On 7 January, at around 17.00, *Discovery II* crossed the Antarctic Circle.[4] After that, they continued for a few hours with little more than a gentle breeze and covered around eight miles per hour. Visibility was clear as the ship gently rolled on her way forward. It made for calm, relaxed progress. In the evening, the ship stopped briefly for their regular echo sounding. However short a time this took it always seemed an interruption to their progress, but it assisted them in their navigation and helped towards a good record for the future. A few hours later, towards midnight, the weather changed. The wind increased, the ship's rolling became more noticeable and gradually they slowed to seven miles per hour. Heavy spray began to fly at them from all sides. The air temperature, already much lower than it had been before the ship reached Antarctic waters, was now 34 degrees Fahrenheit, the sea temperature 35 degrees Fahrenheit.

The wind dropped the next day, and they were able to move forward at almost nine miles an hour. Around midday, at 180° longitude, they crossed the International Date Line, moving to the west. They now logged in a second January 7th. The water temperature noticeably began to drop further. Soon it reached freezing point, then, by mid-evening, it registered 30 degrees Fahrenheit. The chilled water started to freeze, and icebergs appeared on the horizon. In the late afternoon of the new Tuesday, 7 January, six days after leaving Dunedin, they came to the edge of the pack ice. At 17.40, in 66° 48′ S 178° 54′ W, the outstanding challenge of their journey lay immediately ahead of them.

Hill had received ample advice before he set out, including a telegram sent from the Discovery Committee on 19 December, as they were beginning to finalise their plans. This clearly summarised his instructions. Once they were through the pack, they should cover whatever remained of the Ross Sea for them to reach the Bay of Whales. There they would start their air search for Ellsworth and

Kenyon and continue to scan the terrain until 1 February. Inevitably, everything would depend on the weather conditions. If these remained adequate for further aerial search, they could continue for a short while after that. Then, Hill should ensure that enough supplies were left in the depot at Little America for any future comers and, whatever the conditions, be sure to leave the Bay of Whales no later than 15 February. After that, they would leave those on board *Wyatt Earp* to continue the search, even, if necessary, through the Antarctic winter.[5]

Choices open to Hill at the Bay of Whales did not concern him here; these would remain for another time. Now, as *Discovery II* approached the edge of the Ross Sea pack ice, the question was whether or not this was the moment to enter it. Both the Discovery Committee and John King Davis had seen this to be a decision point, and in their advice the Committee had given Hill two clear alternatives. Either he should stay at the ice edge to await the arrival of *Wyatt Earp*, or immediately make his way into the pack. Davis had added the possibility of steaming along the edge of the pack to find more favourable conditions if required. Hill should not, however, approach the coast of Victoria Land where there was greater risk of accumulating ice.[6] As they made their final approach to the much-feared pack, Hill smiled ruefully as he remembered the rather obvious, though well-intentioned, advice to choose whether or not to enter the ice which lay ahead. To Hill the decision was easy.

Had *Wyatt Earp* been present, the wooden ship might have helped *Discovery II*'s progress by pushing ahead through the ice in the manner long established, leaving open water astern of her where the second boat could follow. However, Wilkins' ship was still thought to be somewhere off Marie Byrd Land, over 1500 miles away. There was little likelihood of *Wyatt Earp* joining them any time soon. Wilkins had his own programme to follow. He neither would, nor should, abandon lightly his main aim of reaching the Bay of Whales to rescue their own aviators.

A much more crucial consideration for Hill lay in the undoubted risk that, were *Discovery II* to linger at the ice edge, there would always be the chance of her being pushed against the ice should the wind change, a swell develop, or a blanket of snow come to block their vision. It was far better to find a more sheltered position and to take a safer course through a lead in the ice where, with good fortune, they would be protected from the swell.[7] Hill had no doubt about the right decision in the present circumstances.

It was a beautiful evening, with little wind and hardly any swell at the time, but there was no knowing when either might increase. The ice looked open, and all might have been well if they had stayed. There would, however, be no advantage in lingering, and Hill took no time in settling on his course of action. Instantly he chose the second alternative and began to enter the ice. In compliance with the Ice Regulations, he wrote concisely in the logbook:

> Vessel entered pack ice in order to make passage to the south in search of missing airmen Ellsworth and Kenyon. Speed various whilst navigating in loose pack ice.

From then on, Hill always wrote up the logbook himself or countersigned it. For several days he and Walker stayed up all night. They took alternate watches, neither of them taking any sleep for three days.

They reduced speed by almost two-thirds, to a little over 30 rpm and went forward keeping their speed to no more than three miles per hour, varying it as the moment required.[8] Soon they started the zigzag movement necessary to follow the channels of open water as they appeared ahead of them in the pack. Lumps of ice, smaller at first then increasing in size, fell back from the ship's bow as she moved slowly forward. Shoals of krill moved rapidly away from their bow. Astern they spotted orca, often called killer whales, rising and plunging in their gentle wake.[9] From time to time they saw seals lying out on an ice floe. Occasionally a blue whale blew nearby and surfaced briefly. The sea was calm and only faint ripples appeared on the surface. The pale light of the sky fused with the greyer shades of the water and added to the peace that now enveloped them.

Later, in a particularly large pool of water, they had a most unusual sight: in the still of the evening and the sea almost glass-like, krill were breaking the surface of the water, sometimes jumping well out of it. The fluttering on the surface of the water gave the impression of fine shot being sprinkled on it. Only the perfect stillness of air and water allowed this phenomenon to be visible. Had the water been moving, they might never have been seen it.[10] The first evening was a gentle and beautiful introduction to the monochrome world of Ross Sea pack, enclosed as they were in a new and, to most of the Australians at least, a very different world of white.

Chapter 21 Following the Leads

Now, for the first time on this journey, at 66° S and not long after the Antarctic mid-summer, they had permanent daylight. While the sailors remained focused on manoeuvring the ship through the ice, the airmen made good use of this extra light and the still weather to set up the aeroplanes. They had at first intended to wait until they reached the Bay of Whales before completing their preparations. Now they decided to take advantage of the daylight and the good weather they were enjoying. There was only the lightest of swells to disturb them. They realised that this was an opportunity not to be missed, and that they would have less to do once they decided to fly. They moved swiftly and managed to put the main wings and the tail on the Moth. By the time they stopped, around 01.30, there was little left for them to do.[1]

Gradually the ice floes increased in both numbers and size. Before long, in the early hours of the next morning, they were surrounded by walls of ice which rose to as much as eight, sometimes ten, feet above the water and frequently stretched unbroken for several hundred yards ahead of them. It was an awe-inspiring scene. Once they had completed their work on the Moth, the Australians spent hours simply gazing at the shapes and forms around them. For Douglas, who had flown for Mawson on the BANZARE expedition with the first *Discovery,* ice such as this was not new. For the other airmen, it was the first sight of such formations.[2]

The beauty was overwhelming, but everyone knew that there were hidden dangers. So far, the ice floes, even the small bergs, could not be compared with the size of the floes and bergs some of those on board had encountered almost four years before in the Weddell Sea. Nevertheless, small though they were, even these could still cause immense damage. Their shelves of ice could protrude just below the water line and pierce the sides of the ship in a moment, or easily snag the propeller. When *Discovery II* struck ice, as by this stage she did frequently, there was always an intake of breath, for they could never be sure what damage might occur.

In no time Hill was repaid for the tense times he had experienced back in Melbourne with Captain John King Davis. At first, Davis had been adamant that *Discovery II* was incapable of working through the Ross Sea pack without serious damage to her hull. Later, however, he had yielded to Hill's insistence and to the evidence given by the Lloyd's representative. The 'Ice Notes' which Davis had sent to Hill when *Discovery II* was about to leave Melbourne now came into their own, with advice which would prove invaluable. These briefings on the most suitable way to manage a passage through the pack were based on Davis' own six voyages to the Antarctic.

In these notes Davis adopted a far more mellow and helpful tone than he had shown when the two first met. With *Discovery II*'s ice strengthening, he wrote, the conditions of the Antarctic summer and, above all, with the right approach in navigation, the journey could be manageable. But at every point, Davis emphasised, Hill would have to take precautions. His own experience was based on voyages in wooden ships. *Discovery II* had many advantages over the old-style ship, especially speed and far greater manoeuvrability. Also, the coming season, high summer in the Antarctic, would help considerably once there had been some ice melt. Even so, Davis wrote, *Discovery II*, compared with the wooden hulls, remained fragile for the intended journey. Though *Discovery II* had some degree of ice-strengthening, Hill was to be careful, and he must at any point be prepared to draw back if necessary.

Davis advised Hill never to force his way through ice by adopting the old practice known as 'butting the pack'. Far better, he wrote, to give the pack a gentle nudge, at a speed not exceeding two and a half to three miles an hour. The ship should only try to proceed once the pack could be eased open, the bow pushing slowly into a suitable fissure. Only then, Davis wrote, should Hill consider pressing gently ahead. Davis stressed the 'gently'. If the pack closed round the ship the wise course would be to remain stationary and await a change of wind.[3]

Davis emphasised two areas they should take care to avoid. One was Scott Island, a volcanic outcrop, at 67° 22.42′ S 179° 54.42′ W. The other was just over 300 miles to its southwest, the region near Cape Adare in Victoria Land, on the Antarctic continent itself, at 71° S 170° E. Hill should not consider going closer than 50 miles to any part of Victoria Land. The prevailing wind blew the pack ice steadily onto both Scott Island and Cape Adare causing it to mass into dense, impenetrable mounds. A ship trapped in such ice would find it impossible to escape.[4] Such had been the fate of *Southern Cross*, which was locked in the ice there for six weeks, with Carsten Borchgrevink, the Norwegian explorer who would subsequently be the first to land on the Antarctic Continent. Davis did not mention, at least in his Ice Notes, the fate of *Aurora*, whom he himself had sailed back to the Antarctic once she had been repaired, to rescue Shackleton's men. He might well have raised the point. *Aurora's* fate was a warning to them all. To spend over a year locked in pack ice as had that ship, drifting steadily northwards at the pace of the floes to which she was captive, was the last thing for which anyone would wish.[5]

Discovery II continued steadily the next day, January 8, never exceeding three miles per hour. The comment in the logbook was always 'navigating through loose pack ice. Damage if any unknown'. This was a phrase to be repeated many times during the following days. But overall, the pack remained quite open, and for the time being they made good progress. Compensations came their way and helped. The buffer of snow lay quite deep on the floes and gave them protection, as it softened any impact the ship made with them. There were no icebergs in sight. At most, the ice was still about eight feet thick, and the weather good. It was a silent world, both vast and humbling, and, so far, *Discovery II* was cutting her way through the ice with no great problem.

Throughout the morning of 8th, *Discovery II* made the same steady progress as before. From time to time, the men saw seals stretched out on ice floes. Occasionally a blue whale spouted ahead of them. One moment they would be steaming in open water, well clear of pack ice, the next they would have to navigate slowly again, going carefully because of the loose ice surrounding them. Around midday, they reached 67° S and soon passed safely to the west of Scott Island, leaving one of the places of which Davis had warned well over fifty miles away, and clearing any of the dangerous pack ice that might have massed there. The island made a good marker and staging post on the journey but, clear of ice though that stretch had been, it was crucial to remain wary. Conditions changed constantly.

In the afternoon the ice was a little closer. Hill set a maximum speed of 30 revolutions through this belt of pack until they reached a large lead at 17.00. They increased their speed according to the size of the leads that followed until, to their surprise, at around 21.00 the same evening, they came to an unusually large expanse of open water in 68° 24′ S 178° 0.15″ W. It seemed for a moment that they might have passed the bulk of the pack ice, and Hill called to move a little faster. But their hope was short-lived. Only two hours later they met more ice, and soon after midnight Hill was forced to slow the ship down once more to almost half speed.

That evening, until then, they had still been making good progress and could steam at full speed. The weather was light, and they had reached 68° 20′ S 179° 30′ W. At this point they heard from Wilkins that *Wyatt Earp* was still off the coast of Marie Byrd Land at around the same latitude as they were, but at 141° W, passing along the Ruppert coast at the foot of the Rockefeller Plateau. This was considerably further to the east than their own position and could give Wilkins a considerable advantage if they continued at their current pace.

Wilkins was undoubtedly getting closer to their joint goal but, like *Discovery II,* he still had some distance to cover. He could well encounter heavy ice to the north-east of the Ross Sea, yet one more area well-known for pack ice to linger well into the warmer season.[6] The concern of each of the rescue teams was to get help to the American explorers, but it was too soon to say which would be the first to reach the Bay of Whales.

The next day, 9 January, one week into their voyage south, proved more testing than any they had experienced so far. Occasionally they found stretches of open water but these welcome areas, as before, alternated with tightly packed floes which rose up to ten feet high. They often presented a solid mass through which *Discovery II* had to force her way. It was always crucial to find the weakest spot on which to put pressure before prising the floes apart. To their relief there were no bergs in sight, and some leads through the ice remained open.

The ship now rarely moved at little more than half speed. The visibility was good, but any attempt to make anything they could call rapid progress had become a challenge. In the more open stretches they met numerous ice streams which, as always, were a call for great caution. Often, these courses of looser ice contained irregularly shaped fragments of bergs and floes driven together by the force of wind and sea swell. Packed densely as they often were, they could prove particularly dangerous. At this point the ice blocks frequently crossed the path of the ship and made the steady advance of the previous day almost impossible.

Discovery II's turns now involved sharper, even more frequent zigzags, as they followed the often tortuous courses of the various leads through these streams. At one time during the morning, they veered from a course of 185° to 218°, then back to 190° in under two hours. In the course of a few hours, they repeated the zigzag well over a dozen times. That day they averaged three miles per hour, just half the speed of the previous day and were hard pressed to cover even 13 miles. The bumps and crashes against the ice sounded out ever more frequently, each impact louder than the one before. Every time they struck ice the men still held their breath as they wondered what damage they might have sustained this time. The sound was, as always, disconcerting, especially to those below decks, where every bump seemed like a pistol shot. But everyone remained confident in the calls from their Captain. Watching from the bridge, Douglas was impressed with the way *Discovery II* took each impact. He described the sound, with considerable understatement, as 'unpleasant', saying that it gave the feeling that if the ship pushed too hard, the plates would be 'stove-in'.[7]

As Hill knew only too well, this could undoubtedly be the case. His caution was well founded. The method of opening a lead by a steady approach, followed by gentle but firm pressure, as John King Davis had recommended, was wise and the ship seemed to cope well. In the ship's favour, the wind had decreased and remained a light Force 2 for most of the day. Their speed throughout this time was controlled by the number of floes already in their path. Their aim remained to avoid, where possible, having even more ice streams forced upon them. All the time the weather was becoming steadily colder as the air temperature dropped to 29 degrees Fahrenheit and the sea temperature to 30 degrees Fahrenheit.

In mid-afternoon on 9 January around 16.00, they came to heavier streams than before.[8] Though these were still fairly loose and relatively free moving, they made for even slower progress than ever. The ice began to close in on them once more, and the weather to deteriorate. Clouds appeared and the wind freshened from the south-east. bringing with it a touch of light snow. Once more Hill gave orders to go no faster than 30 revolutions. With some considerable effort, they came to a large lead in the ice. It was by then around 17.00. Later that day, at 21.15, they were steaming ahead at full speed in open water, completely clear of pack ice at 68° 2′ S. Then, by 23.04, they met loose pack ice ahead of them and were forced to return to slow movement through the lumps. Once again they varied their speed accordingly, as they came first to one lead, then to another.

Discovery II's trials were far from over. At 08.00 on 10 January, at 71° 0.37′ S 177° 42′ W, she met even more pack ice. In a moment this turned out to be the heaviest they had encountered so far. Added to this there were several icebergs in sight.[9] From this moment on, their progress south, so far so rapid, took a serious turn for the worse.[10]

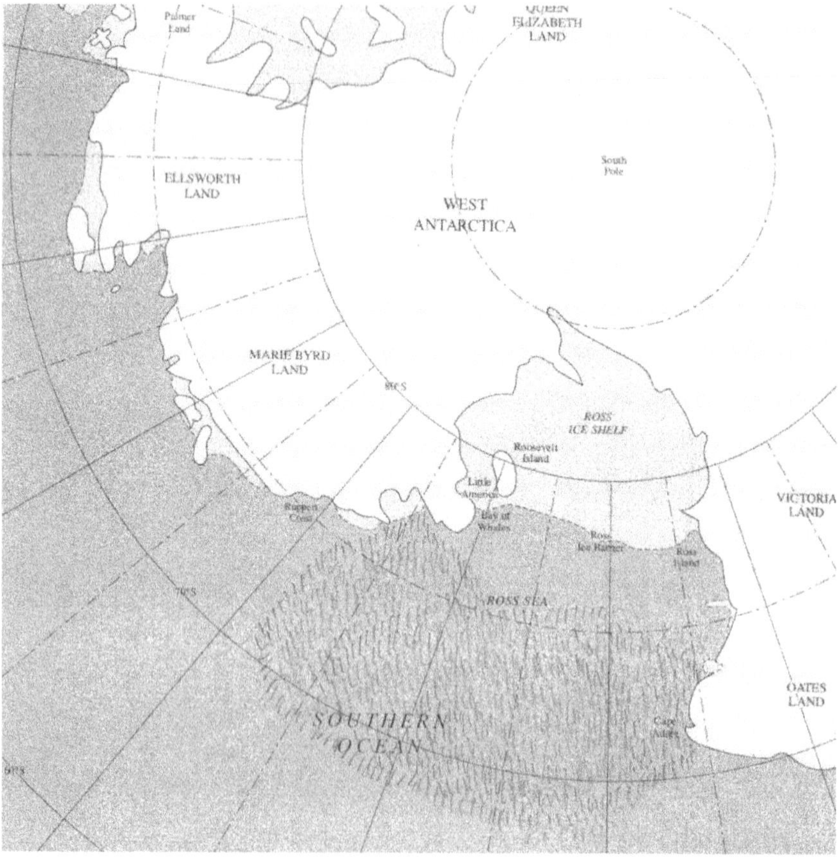

Map 3 Ruppert Coast to Cape Adare, West Antarctica, showing Ross Island, Ross Ice Shelf, Bay of Whales and Little America, and showing general area of Ross Sea pack ice.

Chapter 22 Ice Bound

In the early hours of Friday, 10 January, two and a half days into the pack ice, they managed to put on more speed, but by 08.00 *Discovery II* was surrounded by the heaviest pack ice she had met so far. At 177° 58′ W Hill had brought the ship as far to the west as they should go if they were to enter the Ross Sea without encountering massed ice to its south-west, such as they would meet near Cape Crozier on Ross Island. Equally importantly, were they to continue on this line, they would be giving themselves unnecessary extra mileage eastward along the Ross Ice Barrier, once they reached it, in order to arrive at their final destination. Hill now settled first on a southward course at 180°. Later, at 13.00, he turned to 190° so that they could eventually work their way eastward when crossing the open stretches of the Ross Sea itself. Using as much as possible the leads which led in an eastward direction, *Discovery II* pursued this zigzag course steadily until midday. But then the floes became even more heavily packed, the ship's contact with them ever more frequent. They soon became locked into the ice with no way forward.

As he saw this scenario developing during the morning, Hill called for both anchors to be lifted clear of danger. If left in place, these could easily be caught in the ice. Again, if forced in the wrong direction while embedded in ice, they might rip the outboard face of the hawse pipe through which the anchor chain ran. It was an essential precaution, which he had learned from *Discovery II*'s experience in the Weddell Sea in 1932. Hill called for the ice anchors to be laid ready on deck as an alternative.

It was 16.00 before they could move forward again. In the next four and a half hours they made only half a mile. Then conditions deteriorated even further. The ice steadily accumulated around the ship and appeared ahead of them in daunting fashion. Wherever they hoped to turn, yet another ice wall appeared ahead. To their advantage there was still hardly any wind. The pressure remained steady, but the water temperature was dropping. At 22.30 they brought the ice anchors into use and with them carefully moored the ship to a large floe. There Hill decided to lie to for the night. They were by then at 70° 30′ S, with 450 miles still to go.

By this stage everyone was longing for rest. It had been a particularly exhausting day. Hill and Walker had remained on the bridge since they first entered the ice three days before, and their eyes were now heavy with sleep. Others too were suffering from lack of sleep. With the influx of extra people for the voyage, many of the ship's staff had been forced to adjust their normal sleeping arrangements. Several were sharing cabins or had moved elsewhere. Saunders, the photographer, had yielded his cabin to two of the airmen, and he records finding the bridge, where he had been relocated, such a hive of activity that he hardly slept at all.[1] Some, who had already started the exhausting task of pushing the ice from the ship with poles, were aching from the unaccustomed exercise. That day, the airmen had set up some staging on either side of the stern to enable them to fix some of the struts they required to deal with the Wapiti. As there was so little wind, it had been pleasant working on deck, but they too now welcomed the rest. For others the incessant sound of the fusillade from the clashes with ice allowed for little sleep. Inevitably a degree of exhaustion had begun to set in. There were several bergs in sight, each one of them potentially dangerous, and the men on duty kept careful watch throughout the night. For the time being, *Discovery II* remained moored to the same ice floe. Most of those who were off duty tried to get some sleep, while Hill and Walker took turns to remain on duty.

They were now at 71° 37′ S 177° 42′ W, a point where, in other years, they might well have come to the southernmost part of the ice belt. However, the next morning, Saturday, 11 January, the ice, now much heavier and more compact, was even more tightly packed around them. A wooden ship might well have withstood the pressure if more ice were to be forced upon it. But by this point *Discovery II* was approaching the limit of what she could stand.

It was generally accepted that once a ship had navigated the pack ice at the entrance, the Ross Sea itself would be free of ice. They still had to cover 500 miles to reach the Bay of Whales.[2] Hill felt they could not be far from open water and would have liked this to be a suitable moment for the airmen to go up in the Moth for a reconnaissance flight and circle round to survey the extent and the condition of the ice ahead of them. They might also be able to tell more exactly how near they were to open water. Above all, the airmen could see what leads might present themselves. But there was no water-space for the seaplane to take off.[3] For this Douglas would have required at least 400 yards of clear water up-wind. Instead, as an alternative, at 05.30 on 11 January Hill followed the time-honoured custom and sent an officer aloft to report conditions.

Lieutenant Marchesi, who had presented himself for the task, surveyed the scene from the top of the mast, scanning the ice in every direction. What he saw was not promising. From 60 feet above sea level, Marchesi saw no sign of any darkening of the sky that might indicate open water ahead. All he reported was the glare of ice in every direction. Hill had been hoping that Marchesi might somewhere spot a sign of 'water sky', with clouds looking grey on the underside and suggesting dark water rather than ice beneath them. In that case he would have felt some encouragement and turned towards it. With no sign of water sky, he found the situation, in his carefully understated words, 'perplexing'. He even considered turning to the north should water sky be visible in that direction. But nowhere could they detect any darkening of the sky whatsoever. *Discovery II* was, in effect, truly ice bound.

Hill remembered clearly that some areas of the pack had gained notoriety as trouble spots to be avoided. *Discovery II* was well distant both from Cape Adare, where *Southern Cross* had been delayed for six weeks, and from Cape Evans where *Aurora,* the ship for Shackleton's other party, had initially been trapped in ice before being pushed northward for so long. There may have been no danger of an exact repetition of those episodes, but the accounts inevitably surfaced in Hill's mind as he now thought of the possibility of ice pressing in on them even further. These were tense moments. He knew they had to try to keep moving in whatever direction they could manage, if they too were to avoid being locked into the ice for any length of time.

Careful study of the available records showed that most years, unless ice was unusual that year, they would reach open sea to the South, if not at about 72° S, then without doubt at 73°. They were then lying at 71° S. Hill calculated that the clear water they wanted to reach was only 83 miles away. To retrace their route would have meant 311 miles before they reached open sea to the north. He decided to continue and was amply encouraged by the enthusiasm of the entire ship's company as they cheered in support.

They cast off to face another chequered day. By 08.00 they stopped in the pack, ice axes laid out ready for them to anchor in the ice at any time, but they moved on again in under an hour. They proceeded through loose pack ice until 11.05, put out the ice anchors, and stopped in the pack for 20 minutes. In the next four and a half hours they advanced just one quarter of a mile, choosing the most likely lead from the few which presented themselves, in the hope that it might be the one most able to take them to open sea.

At this point, the 'poling party' came fully into action and proved themselves a vital element of their safety.[4] For this, everyone available, scientific officers, airmen and seamen together, stationed themselves aft with poles. At a system of whistles, an officer indicated to Hill whether he should manoeuvre ahead or astern. If *Discovery II* failed to move in relation to the ice, the poling party simply thrust harder, using every bit of force they could muster, as they struggled to move the floes away from the rudder and the propeller. These parts were especially vulnerable, as Hill well remembered from the episodes in the Weddell Sea. On that occasion, not only was the ship holed to forward, but the rudder was bent. This must not happen again. The risks were far too great. Such a disaster must be avoided at all costs. Next, one of the officers in the poling party signalled to the bridge whether or not they had succeeded in protecting the rudder. It was a delicate task, yet it required considerable force as they leant on the heavy poles to push away the floes and create space between the jagged ice and the stern of the ship. Deacon reckoned that they sometimes exerted a push to their side of nearly one third of a ton. It was not much, he acknowledged, but, with no wind to deflect them, it was surprising how well even this pressure enabled them to push the ship's stern round.[5] Hard work though it was, it was an essential part of the operation and, without doubt, it gave the scientists something very different to do from the continuous stations and laboratory work in which they would have been engaged on a normal cruise.

In addition to the poling party, who were working to protect the rudder and propeller, other people played an important part. Those pumping oil into the tanks left the drums in the after hold full to keep the stern as low as possible in the water. They also removed the storage drums still on deck to any space available beside them. This action kept the weight astern as low in the water as possible, increasing the after-draught and providing additional protection for the rudder.[6]

Aft, outboard platforms had been rigged to assist the airmen when they worked on their aeroplane. These platforms now proved invaluable for a different purpose. They projected about 16 feet over the sides of the ship.[7] Anyone lying flat on them could watch the propeller and the rudder from close quarters. From that position someone would shout the alert should they find themselves too near to any ice.[8]

Forward, Saunders had already spent ice-cold hours in a Bo'sun's chair, which he had slung below the bow. That day he perched on it, taking photographs, acting as forward lookout and watching for the otherwise invisible ice spurs which might appear ahead. It was a role which carried its own risks, as these protrusions of ice would appear suddenly at his own level ahead of the ship, and on many occasions Saunders had to leap back inboard to avoid them.[9] It was also bitterly cold for him as he lay there, but his work too proved to be of invaluable help.

Morale on board ship remained high. Around midday news came that Wilkins was still working westwards, skirting the pack off the coast of Marie Byrd Land at about 66° 20′ S. It was known that the pack ice along the Pacific coast between 75° and 150° ice floes rarely broke up and remain locked in that area for several years.[10] Now it seemed that *Wyatt Earp* was being pushed northwards by heavy ice at the entrance the Ross Sea.[11] *Discovery II* might yet reach her goal before the American ship.

Slowly they edged forward. As was always their practice now, they never rammed the floes at full speed to force them open in the way a wooden ship might have done. Instead, they continued their new practice of working slowly through the cracks, nudging them open bit by bit and pushing gently at what might be the weakest part. On occasions when no cracks were visible Hill would bring the ship gently head-on to the floe. Once the ship made contact with the floe it was safe to increase speed until the pressure created an opening in the ice. When they achieved this, they could manoeuvre gently through the slowly widening space. There was always a sense of relief when they first sensed a crack appear in the ice. Fresh tension would then follow as they hoped that there were no jagged outcrops of ice beneath the water surface. Many of the floes were extensive. Captain Scott had referred to floes as large as football fields, or even larger, and so the floes seemed now to the men in *Discovery II*, the sides of the ice rising sheer, often up to ten feet above the water.[12] Fortunately, however, the impact with the floes was often cushioned by the overhanging buffer of soft powder snow which acted like a fender for the ship's side and prevented contact with the sharp ice spurs beneath the floe's overhanging edge. As always, it was these numerous, unquantifiable outcrops below sea level which caused the greatest concern.

By midday on 11 January, sightings showed that they had returned to 71° 35′ S 177° 34′ E. Although they had moved about four miles through the ice itself, they were two miles to the north of their position of just 24 hours before, at 71° 03′ S 178° W. Most probably, a current had worked against them by pushing the ice northwards. It was always a game of suspense where their mood swung from hope and optimism to frustration and deep concern. This setback proved particularly deflating and was all the more disappointing in view of their efforts the previous day.

Discouraging though this was in terms of progress gained or lost, a few things went well. Douglas and the other airmen took full advantage of the relative comfort the slow progress rendered necessary and worked on preparations for the Wapiti. They opened the boxes containing its skis, then spent much of the day cleaning them and stowing all the small parts below decks. They also prepared the ski runners of both planes to have them ready whenever they might be needed, two pairs for each seaplane.[13] The airmen then stowed one set below deck and made adjustments to the other two, leaving them to hand for any emergency.

Meanwhile, the many Adélie penguins who cavorted on the ice around the ship caused great amusement. In the early afternoon, when the ship was stationary beside a particularly extensive floe, Douglas joined them and skied over a long area of the snow-covered ice. His intention was to see if it could be used as a runway for the Moth. Ideally, they needed at least 300 yards for safe take-off. This stretch of snow on ice, 200 yards wide, would have offered them up to 350 yards in length as runway.[14] It might be suitable in an emergency, Douglas concluded, but it had numerous ridges. The men would have had to flatten these before he could take the Moth over it, a task that was certainly not worthwhile in their present situation. He and Hill decided to wait in the hope that they would find an easier stretch, even if it meant delaying until the next day.

At 17.00 there was still no sign of the ice offering any leads. By the end of the day, they had only advanced about one and a half miles. During the day they had simply experienced the lightest of breezes, with a complete calm throughout the afternoon. Their only chance lay in a fresher breeze from the south-east to loosen the chunks of ice massed around them.

Chapter 23 HELP FROM THE AIR

At 17.00 on 11 January, there was still no sign of a way opening up through the ice. Hill continued to hope for air movement which might shift some of the ice piled up around them. However, had the breeze freshened, another problem would have presented itself. This lay in the way *Discovery II* responded to the wheel. In an ordinary vessel the bow moved to starboard when the engines were put astern. With *Discovery II*, the stern tended to move in the direction of the wind, whether the wind was on the starboard or the port quarter. The wind was barely noticeable on the surface of the water but had slightly increased above the water level. Light though the breeze might be, even then on many occasions it was difficult to anticipate how the stern would move.[1] Clearly, in the tight space available, any move towards hidden ice spurs could remove all the benefits the poling party had gained for them during the day. Hill decided to lie to for another night and await better conditions. Once again, they anchored to an ice floe for the night and remained stationary for nearly 12 hours. It was during this time that the scientist, Ommanney, suddenly saw the unmistakeable signs of a fin whale blowing in a small opening not far from the ship. It was a clear sign they were approaching more open water.[2]

They might have stayed longer had not two icebergs drifted through the pack ice in the early hours of 12 January. Bergs were always to be avoided in pack ice if at all possible.[3] Hill recalled vividly the sight of a berg driven by the wind, moving relentlessly towards them as they moved northward from the Weddell Sea in 1932. A moving iceberg was unassailable. With one now approaching them, the only way for them, Hill knew well, was to continue in their efforts to work their way through the pack, however difficult this might be. The alarm was immediately raised.

At 04.00 they brought in the ice anchors and 25 minutes later moved forward. After two hours of tedious going, they reached 71° 38′ S 177° 34° W. By then conditions had begun to ease, and by 08.00 they at last reached a pool of open water.

This was the first pool they had met in the last 24 hours which was large enough for the Moth to take off with safety. The water surface was only mildly rippled and, although the weather was far from perfect, Douglas offered to attempt a reconnaissance flight. Ideally the airman would have liked conditions where, with a clear sky, he could rise to as much as 6,000 to 8,000 feet above the ship. From that height he would be able to clearly see pack ice conditions around them for at least 70 miles.[4] On this occasion the sky was cloudy. He would have been satisfied if he could climb to just 2,000 feet, from where he would be able to see a minimum of 45 miles. In spite of the limitations, he knew he would face, Douglas was still prepared to inspect the ice from the air. He assured his Captain that he would limit himself to circling the ship, rather than venture over ice. Hill accepted the airman's offer. A view for even 45 miles was far better than the 13 miles a man could have surveyed from the crow's nest at the top of their mast, a mere 60 feet above sea level.

Before taking off, Douglas ran the seaplane's engine for about 15 minutes to warm it and do an initial test. Hill arranged for the ship to be manoeuvred so that the starboard side was in the lee of the wind, then he stopped the ship's engine. Once Douglas had climbed into his cockpit, men hoisted the Moth on the starboard side of the ship. and swung it well clear of the ship's side. They then lowered it gently onto the water, with its nose facing inboard. They let the ship's pram dinghy down at the same time and used this as a possible buffer to keep the seaplane away from *Discovery II*'s side. The motorboat then towed the plane out into more open space and waited there in case the Moth made an unexpected, forced landing.[5]

Douglas manoeuvred the plane for a short while, then took off, unaccompanied, at 10.15. He took the Moth up in circles as high as he could. To his great disappointment the cloud cover meant that he could climb no higher than between 1,000 and 1,200 feet, considerably less than reach the minimum 2,000 feet for which he had hoped.[6] He returned 30 minutes later and landed to leeward of the ship. Then, after careful manoeuvring, the ship's motorboat towed the Moth back to the ship with Douglas still seated inside. Men attached lines from the ship to the plane, and Douglas was aboard ship a quarter of an hour later. He immediately reported to Hill what he had been able to see.

Even though his survey was limited, it still provided valuable information. From his cockpit, at 71° 45′ S 178° 04′ W, Douglas had enjoyed a view of over 40 miles. He realised that the ice further ahead of the ship was softer, and presented more open ways between the floes, seemingly leading to what looked like clearer water between 30 to 35 miles ahead.[7] Douglas' aerial reconnaissance gave Hill hope at last of a manageable way out of the ice.

For the rest of that day, they continued through loose pack ice. Visibility was good so Hill could vary his speed according to whatever surrounded them. They progressed well until 17.00 when some old, very rugged floes blocked their way completely. As a moderate wind was blowing, Hill decided to wait rather than run the risk of these being blown onto the ship. It was a wise move, and fortunately they did not have to wait for long. An hour later the wind had made an opening in the floes, and they proceeded cautiously for two hours. By 20.00 Hill felt confident enough to take alternate four-hour watches.

The wind was gentle, and at midnight there was a brief spell when they recorded complete calm. As always, what unnerved them was never knowing what damage they were sustaining from the frequent clashes with the loose ice. Nor had they any real clue which direction would be the safest for them to move next as they followed their zigzag course southward.

The trial was not yet over. Towards 08.00 the next day, Monday, 13 January, the pack thickened yet again. For a while more lumps of loose ice streamed across their path. Old, well-worn and jagged, their appearance was by now a familiar sight, but, with their ragged and unpredictable mix, ice streams presented a particular hazard and officers had to watch them closely.

This was, however, not a time for Hill to be deflected from his purpose, and *Discovery II* continued in the same way throughout the morning of 13 January. Soon after midday they reached another pool of water large enough for Douglas to feel it reasonable to fly the Moth for a second time. The seamen launched the motorboat and, in spite of occasional snow squalls and a wind speed of around 15 miles an hour, it towed Douglas out for another reconnaissance flight. Murdoch accompanied him this time and Douglas again reached a height of 1200 feet. On this occasion the return proved more difficult, made the harder as they were in a machine as small and light as the Moth.

Conditions had been difficult at every point for a flight in their type of plane. The air temperature was as low as 30 degrees Fahrenheit, it took time to start the engine, and visibility was disappointingly poor. During the hour the flight had taken, the ice surrounding the pool of water had closed considerably. In ideal conditions their landing strip should have been free of any obstruction, however small, but now Douglas and Murdoch found that numerous old bits of ice lay in the way. Added to that, the wind speed had increased to 18 miles per hour and the sea had become quite choppy. It was generally recommended that the plane should be brought in on the lee side, with the ship as near stationary as possible. Now, however, the wind was on the wrong side for them to return aboard.[8] Douglas asked for Murdoch to be taken off the seaplane in the motorboat. He then signalled for the ship to be turned round. This Hill instantly put into effect. It was impossible, contrary to the usual requirement and to common sense, to keep *Discovery II* stationary for the manoeuvre required to hoist the Moth out of the water. Even though the plane was taken aboard on the lee side, it proved a difficult exercise.[9] These were testing conditions for the seaplane and seriously challenged both the airmen and the ship's own crew alike. Hill found these moments were even more tense than when he had watched the small, yellow machine circling above them.

It took a full half hour after they had first touched down to bring the Moth inboard, but when at last the two airmen were back on board, Douglas gave Hill news that proved to be of the utmost value. As he and Murdoch had strained their eyes towards the West, Douglas said, all they could see was pack ice, dense and unbroken as far as the horizon. Once again, their view had been limited because of the cloud cover. This time their view was restricted to a little over 40 miles. Even so, it was far further than any man would have seen making his observation from the top of the mast. Below them, Douglas and Murdoch had observed an intricate pattern of ice and water, but the sight brought a smile of relief to their faces. Just within their view, at the limit of their horizon, lay an open pool of water about 28 miles to the South.[10] While the rest of the scene was daunting, that outlook was without doubt encouraging.

It was a humbling moment for them when they looked down from the Moth and saw *Discovery II* just a solitary, tiny and isolated form in the vast expanse of white, a vulnerable speck on the eastern edge of a silent territory from which, had she entered, there might well have been no escape. Just ahead of the ship the leads in the ice were tight and barely perceptible, but in the distance southwards, they could make out several good channels between the floes. If *Discovery II* could work her way towards these leads, Douglas continued, she should find open water beyond.[11] This was encouraging news for Hill to report to the Discovery Committee and should have brought intense relief to those who followed the movements of their research ship over 10,000 miles away.[12]

Discovery II moved forward at 13.35. The leads Hill now followed during the rest of the afternoon were as tight as any they had faced so far, the zigzags as frequent as they had known, their speed as slow as they had experienced. Hill recorded at least 20 turns in the course of the afternoon and early evening as they continuously changed direction to follow the openings in the pack.

Gradually, however, as Douglas predicted it would, the route became less demanding. To their relief a light wind from the southeast also helped to prevent the blocks of ice from tightening around them. From soon after 22.00, *Discovery II* was able to settle on a course at 185° and from then on she made better progress.

About this time, they heard that *Wyatt Earp* was now free from the ice and making good way. Two days before, the American ship had been firmly stuck in ice and was being pushed northwards following the northward trend of the ice edge. Now, helped by a strong wind from the north-east, that ice surrounding her was breaking up and *Wyatt Earp* was at last able to make her way through more open water.[13] The news caused a stir among the men in *Discovery II* as they realised that Wilkins could now draw ahead. Those who saw the project as a race now wondered if their desire to win had been foiled.[14]

Hill was simply relieved that Douglas' reconnaissance flight had warned them of their own impending danger, and warmly acknowledged their good fortune in being remarkably free of ice. He could so easily have started into the pack to the west, only to become impossibly stuck. Alerted to the dangers, he kept to the eastern edge of the heaviest ice and made his way through cracks and leads until about 15.00. They then reached a lead of 15 miles. From this time onwards they navigated between ice streams and leads until the early hours of 14 January. At 02.50 that day, at 73° 23′ S 179°15′ W, they realised they had at last reached the Ross Sea.

The challenge had been considerable and demanded cool heads, firm resolve and steady confidence. Later, when he wrote his report of the journey through the pack ice, Hill would declare that, though the ice was very heavy, they had never been in any danger. He had been well served by the advice given by John King Davis. He readily appreciated the aid Douglas and Murdoch had given from the air. If they had tried the traditional method of butting the ice, the outcome might have been different. He also fully appreciated the efforts of the poling party which had ensured the safety of the rudder and propeller. Finally, he even acknowledged the encouragement the Committee had sent in their early telegram, with its oblique support for a voyage into ice he would normally have been advised to avoid. At the time they entered the Ross Sea the logbook entry was as brief and succinct as ever:

> at 0250 vessel cleared pack and entered sea clear of ice.

The way now seemed clear for them to sail direct to the Bay of Whales.

Chapter 24 INTO THE BAY OF WHALES

Discovery II emerged from the pack in the early hours of Tuesday, 14 January at 72° S 178° W. Ahead stretched the Ross Sea, ice-free water at last.[1] The belt of pack ice that year had been particularly wide. Most years, ships would reach the southern limit of the pack in January, somewhere between 70° S and 72° S. It was unusual for it to extend further south than that. That year, however, the Antarctic summer season of 1935–1936, *Discovery II* was not clear of ice until she reached 73° 23′ S, around 100 miles further into the Ross Sea than they might have hoped. The last time the ice had extended so far south had been when the *Terra Nova*, the lead ship of Scott's 1911–1913 Antarctic expedition, tried to reach Cape Evans to relieve Scott's Northern party. That was in 1912–13. It was ironic for *Discovery II* to have been the only non-commercial ship in nearly 25 years to encounter a similar stretch of ice. In spite of the unexpected extra distance and inevitable risk, Hill had brought his ship through this unpredictable ice without serious damage in the remarkably short space of seven and a half days.

Scientists, officers and crew had all worked well together. Above all, Eric Douglas' reconnaissance in the Moth had proved invaluable. Without the crucial sightings Douglas was able to report, Hill might well have turned back. Had he pushed forward, his ship could well have been locked into the ice as firmly as the men on the *Terra Nova.* Ultimately, he and his companions would have failed in their attempt to reach the Antarctic mainland and rescue the two Americans. Hill was also conscious that the wind and weather had favoured them. They had not experienced the strong winds that might have blown even more ice upon them than they already encountered. He had good reason to be doubly relieved to be out of the often-deadly ice and into open sea so quickly.

A different problem now emerged. Hill had kept strictly to his routine of telegraphing daily at 12.00 and had been unaware that in the preceding days problems had surfaced with their transmission. A message, cryptic, but understandably anxious, had reached *Discovery II* from London on 13 January – 'please telegraph daily'.[2] This came as a surprise to Hill. He had followed punctiliously his instructions from the Discovery Committee, telegraphing London daily with updates at the appointed time, but it seemed that several of the messages had not reached the Committee in the usual time. News of their greatest difficulties in the ice did not reach London until three days later, when *Discovery II* was well clear of that danger.[3] Likewise, telegrams from London were slower than expected in reaching *Discovery II.* The communication system seemed to be out of phase.

When Hill replied on 14 January from 74° 46′ S 109° 79′ W, with the news that they were by then out of the pack ice, he had no knowledge of the difficulties already experienced. He simply commented that communications were now 'difficult', since to him they were newly encountered.[4] However, more problems were still to come. Excessive swing of signals, which made reception difficult, were above all noticeable when the ship was rolling and in a trough of water.[5] It would take some time fully to explore the reasons for the failure in the wireless telegraphy system.[6] Only when they were leaving Melbourne in early March did it emerge that the new receiver, fitted in 1935, simply was not suitable for a small ship which, with its constant rolling, was frequently in a trough of water.[7] For the meantime it seemed enough for Hill to be able to let the Committee in London know they were through the pack ice and in the Ross Sea.

The Ross Sea was only gently rippled, and at that point there seemed no major hurdle in sight. They had overcome the first challenge, the ice, but the coming days could bring their own difficulties. This was no time to linger. Their next goal took priority – locating Lincoln Ellsworth and Herbert Kenyon. With

320 miles still to cover, they now expected *Discovery II* to reach the Bay of Whales at 78° 5′ S some time the next day. Hill estimated this to be around 1830 hours. They would then start an aerial search, this time, if necessary, using the Wapiti.[8] This was the start of the next phase in their enterprise. Elation and relief gave way to burning curiosity as those in *Discovery II* speculated what the next few days would bring.

Any improvement on their timing now seemed crucial. All minds were on the rescue itself. No-one now mentioned the word 'race' to reach the Bay of Whales before *Wyatt Earp,* but they may still have had in mind what had for some time been seen as such. In the early hours Hill pressed on, cautiously at first, to avoid any last bits of ice that remained, then at greater speed.

Now, with no further ice in sight, Hill called for the ship to turn gradually and make for Discovery Inlet. This was a small bay set into the face of the Ross Ice Barrier, just 100 miles from the Bay of Whales, and the area for which the Discovery Committee had first suggested Hill should aim, at 75° S 130° W. By following the available leads through the ice, *Discovery II* had emerged some 70 miles to its west. By midday, while the ship was already at 179° 59.7′ E, she was only at 74° 45.6′ S. But to have added extra mileage to their passage across the Ross Sea was Hill's least concern. It was an achievement to be out of the pack ice, at whatever point. The outcome, Hill hoped, would be the same. Nor were they unduly troubled that they had to adjust their course at frequent intervals because of their proximity to the Magnetic Pole. From 08.00 that day, Tuesday, 14 January, they moved steadily forward at around nine miles per hour.

Later, the day turned slightly cloudy, but visibility remained clear, and the sea was still only gently rippled in the light wind. In the late afternoon they were surrounded by a stream of ice and stopped briefly to avoid any impact. Just at that point a flurry of hailstones struck the deck, but soon they were again steaming at almost full speed, averaging eight to nine miles an hour. By late afternoon, the weather was once more fine and clear, the visibility the best they had enjoyed for some time. Conditions were near to perfect, and they covered over 70 miles that day.

These were peaceful hours after the tension they had experienced in the previous few days. Meanwhile, the ship's deck once again became a scene of great activity. *Discovery II*'s men prepared for their search for Ellsworth from the Bay of Whales, and the Australian airmen checked that everything was set to enable an aerial search should they not immediately find the aviators.[9]

The next day, Wednesday, 15 January, around midday, they at last saw on the underside of the clouds on the horizon the white glare of ice blink and realised that the Ross Ice Barrier, later to be known as the Ross Ice Shelf, lay less than 30 miles ahead of them. For the uninitiated, this could have been taken as land. In reality, it was a vast shelf of floating ice which stretched for over 500 miles further south to the true Antarctic land mass. At 16.25 Hill turned to steer on a bearing of 184° and set *Discovery II* to steam parallel to the Barrier, about two miles out to sea, on the last leg of their journey towards the Bay of Whales. It was a pleasant time on deck as the airmen and others made final preparations for any rescue flights that might follow.

For some time the conditions remained good, and they could see 30 miles ahead of the ship. Then, as they cruised along the ice edge, a pronounced mirage seemed to lift icebergs, growlers and bergy bits well clear of the water on the horizon ahead. Gradually visibility changed. On their starboard bow, towards the south-east where they knew the Bay of Whales lay, the ice blink brought an almost dazzling glare to the skyline. Sky, ice and water all seemed to merge into one white curtain ahead of them, and it was difficult to see even as far as the middle distance. Meanwhile, immediately to starboard, the ice of the Barrier towered above them, up to 80 feet high in places. 'Just like the White Cliffs of Dover,' one said. 'Without the grass,' added another drily.[10] All along the Barrier were signs of ice falls from the almost sheer face of the Barrier. As they gazed, men could tell the places where the ice had calved and left deep, vertical crevasses which reached high up, often with ice bridges still intact over the steep drop below.[11] All the time, oblivious to this and to the intruders now sailing through their territory, Minke whales coasted backwards and forwards along the Barrier edge, feeding on the wealth of krill visible all around them.

As the sun moved round towards the south, it became easier to see what lay ahead. Just after 20.00, *Discovery II* came round the West Cape, and half an hour later, at 20.30, they arrived just off the sea ice at the mouth of the Bay of Whales. They had reached the entrance to an inlet which spread some 11 miles wide, with ice walls towering some 80 feet above them. The Bay itself, a deep indentation in the ice barrier, had persisted for a long time. Recently named Roosevelt Island by Richard Byrd, it was the result of an area of higher ground, lying below the ice, at 79° 25′ S 162° W. With its northernmost point around three miles south of the Bay of Whales, around 90 miles long from north to south, and 40 miles wide, Roosevelt Island had originally caused the ice streams which descended to the Shelf to fork around it.[12] As the two ice streams then almost rejoined one another to the north, at the Barrier edge, they created the deep Bay which would become an ideal haven for ships at the start of Antarctic exploration in the region.

From the foot of the Barrier, an ice ledge stretched out into the bay, alongside which more Minke whales cruised in abundance. These whales, with their very noticeable curved and high dorsal fin, and unusually large for their kind, remained seemingly unperturbed by *Discovery II* all the time the ship remained in the Bay. Sometimes they moved in groups of six or seven, at other times in schools of up to 40, all drawn there by the dense swarms of krill, identified as *euphausia crystallorophias*, to be seen in every direction.[13]

....

Here was the Bay of Whales, almost fabled from the history of explorers and the explorations connected with it. First discovered by James Clark Ross in 1842, named by Ernest Shackleton in 1908 for the abundance of whales moving in its waters, and used by

Amundsen as the starting point for his trek to the South Pole in 1911, its history was full of meaning. For the men of *Discovery II*, it provided an almost unreal moment as they gazed on the first of the goals to which Hill and his colleagues had aimed over the last six weeks: a place, normally well beyond the bounds thought suitable for *Discovery II*, and which until recently they had never dreamt of reaching. Now everyone wanted to know one thing only – would they yet achieve the main goal, the rescue of Lincoln Ellsworth and Herbert Hollick-Kenyon?

As many as were free to do so came out on deck and strained their eyes through binoculars to scan the surrounding ice top. Then, in the evening, just after 19.30, less than half an hour after arriving, they caught sight of their first real glimpse of hope, an orange tent on the far side of the Bay, not far beyond them, and two orange streamers flying from posts nearby. At first, they thought there were figures moving just beyond them, on the surface of the ice above. But hope was soon dashed when the movement turned out to be made by a few penguins, at home in their own territory and completely oblivious to the excitement below them.

Orange, Hill knew from Hubert Wilkins' communication, was Ellsworth's signal colour. About this there was no mistake: the tent was undoubtedly orange, as were the streamers which hung close beside it. As to who had erected these markers, Hill could not be sure. They gave hope of life nearby, but the lack of movement was concerning. Speculation broke out once more. Had the Americans set this up to indicate their presence and then returned to what they presumed would have been greater comfort at Little America? Or had this spot, marked out in orange, become their last resting place? Despite this latter possibility, tentative optimism now lifted the spirits of those on board ship.

Hill steamed as near to the position of the tent as possible, blowing the siren and whistle, and firing a series of rockets in the hope that these would attract attention. But no answer came, nor did any sign of human life emerge.[14] The reverberation of their rocket-fire disturbed the air around them, but little else. Were they too late?

Or were Ellsworth and Kenyon back at Richard Byrd's base of Little America, on the Barrier ice four and a half miles to the south-east? Hill and his companions could at least hope that Ellsworth had set up this station, in the form of an orange tent, as a signal to observers at sea. There was one way to check further, the search by air for which they had come ready prepared. The weather was calm, so Hill asked Douglas to fly in over the tent and then make for the charted position where they understood Little America had been established.

Chapter 25 FOUND!

Ahead of *Discovery II* and around her rose a steep wall of ice. There was nowhere in sight to tie the ship, so, looking for the most suitable spot to anchor, Hill took *Discovery II* gently to the mouth of the bay to a point just off the sea ice. At 20.30 the airmen began to prepare the Moth for its next flight. Sailors launched the motorboat so that it could tow the plane to a suitable position for take-off. By now it was two minutes before 21.00. The air temperature was 21° Fahrenheit, far too cold for an immediate start. It took a while for Douglas and Murdoch to warm the engine and to cope with the salt spray that froze on contact with the plane. The airmen made a final run of at least one and a half miles to get the Moth airborne, then climbed steadily to 1,000 feet and set a course for Little America.[1] As the two men started their flight over the ice, they immediately spotted its many ridges and hollows and realised the serious difficulties they would encounter should they decide to land.

Flying conditions presented another problem. The glare from the ice merged with the reflected glare from the clouds, and the airmen could barely make out the true surface of the ice at all. Nor had they a clear view of where they were or where they were going. Conditions could not have been more demanding for Douglas as he tried to keep the Moth on course and maintain a uniform height over the crests of frozen snow and the crevasses to either side of them.

The flight did not last long. To their intense relief and satisfaction, Douglas and Murdoch spotted another streamer about two miles inland. Next, they realised that what they had, from a distance, first taken in the evening light to be the shadows of crevasses, were wireless mast and poles. Then some fabric appeared, stretched out not far from a hut almost buried in the snow. Its colour was the signal orange. Then, as Douglas writes, 'imagine our delight when a man scrambled out from the roof and [started] to wave his arm.' Douglas and Murdoch had not only found the Little America they had been seeking but had also discovered someone who perhaps inhabited the settlement.

After a moment or two Douglas flung a parachute from the Moth with a small bag of provisions. Attached to it was an envelope. From the air he saw the man below them walk on snowshoes over to the bag and bend to pick it up. The figure then returned to what appeared to be a dugout at Little America, with the envelope and the small bag of supplies in his hand. From the entrance he turned to give another wave. The man was none other than Herbert Hollick-Kenyon, Ellsworth's pilot. From inside their hut Kenyon had heard the drone of a plane overhead and had rushed outside in time to wave to the plane and then see something descending by parachute. Once inside the hut, as Ellsworth described later, Kenyon brandished the small piece of paper in his hands. 'From Wilkins, no doubt', Ellsworth thought. 'At last, yet so soon.' He would find the letter written, not by Wilkins as he had expected, but by Lieutenant Hill on *Discovery II*.

Douglas circled round over the dugout and began the return flight to the ship. As he and Murdoch swung round, they spotted to the east what looked like the wing of an aeroplane sticking out of the snow. They had understood that everything from the American expedition the year before had been removed so they assumed that this was the remains of *Polar Star* herself, and a sign that this was the spot where Ellsworth's plane had crashed. They were later to discover that the machine was a Fokker aircraft which had escaped Byrd's clearance.

By this time the visibility had deteriorated further. It was difficult for the fliers to make out the line they should take, and at first their return promised to be every bit as challenging as the outward journey. Fortunately, Douglas was just able to identify from their cockpit a narrow arc of water sky ahead that indicated their position in relation to the Bay of Whales. He then sighted

Discovery II and, keeping the line of the Barrier edge well in sight, flew steadily towards the waiting ship. As he flew, he looked out for the most suitable place from where men from *Discovery II* would be able to cross to Little America, should this become necessary.

It was an exciting moment for all those watching anxiously from the deck of *Discovery II* when they saw the Moth circle round and turn towards the ship. Soon after Douglas touched down on the water, he and Murdoch were hoisted aboard. With the light low and the visibility difficult, it had been, Douglas wrote later, the most taxing flight he had ever taken.[2] Hill's desire to push south as quickly as possible in order to take advantage of the short summer flying season was well justified. With any day they might have delayed, the increasing haze of the Antarctic autumn, already not far off, would have further reduced Douglas and Murdoch's ability to see ahead.

People were eager to crowd round Douglas for details, but the Australian went straight to the bridge to report what he and Murdoch had seen. Hill immediately prepared to send a land party ashore over the sea ice. By 23.15, a little over one and a half hours after Douglas had first set off, a party of five left *Discovery II* to go ashore and meet the person the fliers had sighted. They did not have as far to go as they expected. For, even as they were leaving the ship, the five men spotted a figure coming down the ice slope from the Barrier. This was Kenyon himself who, soon after realising that he had been spotted by the airmen, had set out in their direction.

In spite of the poor light of the late evening, he was already making his way down the steep slope onto the sea ice at the Barrier's edge. Kenyon greeted the men with great enthusiasm. 'I say,' he is reputed to have said, 'it's jolly good of you fellows to drop in on us like this', adding that he was just making sure the ship would not forget about him. His relaxed, matter-of-fact manner would cause much comment in the newspaper accounts to follow, to the point where *Bathos in Excelsis* was the strapline given by one writer in *The Times*.[3]

The shore party from *Discovery II* was eager to go inland straightaway and find Ellsworth. To their disappointment, Kenyon persuaded them all to return to the ship. Ellsworth, he told them, was suffering from blistered feet. The older man would be perfectly safe if he remained overnight in the temporary home they had made at Little America. He himself, however, would be happy to return to *Discovery II* with the ship's party. They all turned back and, together with Kenyon, came aboard *Discovery II* almost an hour after midnight on 16 January. It was as speedy a rescue as anyone could have wished.

Waiting anxiously on board *Discovery II*, Hill was as eager as the original shore party to send some of his people out to the old American settlement and to bring Ellsworth to *Discovery II*, but Kenyon stressed again that Ellsworth would be quite secure at Little America. There was no need for haste. It was simply a problem with Ellsworth's feet that prevented him from joining them that evening. In addition, Kenyon made clear his own need of sleep so that he would feel refreshed when they went out again. A large crack existed between the sea ice and the Barrier, which moved and changed its alignment with every tide. He would have to guide the ship's party over it. Crossing such a gap was no wise task for an unslept person, especially one who would, on this occasion, be responsible for others as well as himself.

With the rest of his mission still to complete, Hill stood *Discovery II* out to sea for what was left of the night. There was, not surprisingly, little sleep on board. Kenyon, though he had asked for the chance to sleep, talked well into the early hours to officers and men. They, in their turn, listened avidly as the pilot recounted the failures, mishaps and ultimate success that had come his and Ellsworth's way since they had left Dundee Island in mid-November almost two months before.

He described how they had run into bad weather about three times on their flight across the Antarctic, landing on each occasion and waiting until the visibility improved. As a result, their plane had run out of fuel on their last leg, around 20 miles from Little

America. He and Ellsworth had made their way there on foot with a sledge and had remained there from early December. It was during one of their earlier stays on the ground as they crossed Antarctica, that they had found that their radio would not pick up signals.

Their time at Little America had been pleasant enough. Ellsworth, however, had left his spectacles for reading on *The Polar Star.* He was very short-sighted and, though there was reading material at Byrd's Base, he was unable to read the books left there. Time must have dragged heavily for him. Not only was the sky relatively bright all night from the midnight sun but he also had to see Kenyon reading volume after volume while unable to enjoy the same pleasure.[4]

All who listened to Kenyon's mild-mannered account were full of admiration for the way he and Ellsworth had managed each stage of their epic journey. They were impressed too by Kenyon's modesty as he described how he managed to take off in the thin air of the plateau. Kenyon gave a remarkably laconic and very short report of the considerable problems and hardships he and Ellsworth had experienced, but he had to repeat the account several times as *Discovery II* people came from their duties or returned to them. Where and when Kenyon took any of his well-deserved sleep that night is not clear.

Having kept *Discovery II* lying to for the rest of the night, not far from the mouth of the Bay of Whales, Hill brought her back into the Bay the morning after, 16 January. At 09.00 Kenyon set out with a party from the ship towards Little America. Richard Walker, the Chief Officer, was at their head. Kenyon guided the men from *Discovery II* over the tidal crack, then returned alone to the ship soon after midday.

Meanwhile, the shore party from *Discovery II* continued on its way. On skis and towing a toboggan, they trekked from the Bay of Whales to the settlement. It would have been difficult to find their way without the guidance of the long line of markers which survived from the time of the last Byrd expedition of 1933–1935. Old masts standing above the snow again indicated the last of the route. Eventually they came to the site and found the small settlement of huts, workshops, radio sheds and living quarters which Richard Byrd had established as a second base for his exploration. Only one Antarctic winter had passed since Byrd left it in early 1935, but the small township was already almost submerged by snow. To the shore party's surprise and delight, they were met by Ellsworth himself, who had emerged from his shelter below the snow and was limping towards them. The American led them to the entrance, and together they all slithered down the entrance shaft to a half-lit but spacious interior. This was where Ellsworth and Kenyon had spent the long weeks since their arrival on 15 December. Originally Byrd's radio station, from where Byrd sent wireless messages to the US, it had provided a welcome haven when they first found it almost five weeks before.

Once the shore party came to a rest outside the hut, the bitter cold struck them with great force. The warmth inside the hut was welcoming. They were also ravenously hungry, having not eaten before setting out. Kenyon had told them they would find plenty of food in the hut and, to Ellsworth's amusement, they devoured the supplies Douglas had dropped the day before. After several hours resting, eating and talking, the party returned to the Bay of Whales with Lincoln Ellsworth himself.

The return journey took several hours. The surface of the snow was reasonable enough for them to trek over it, but visibility was poor, and the sledge they pulled now seemed very heavy. Above all, they had to take their time because Ellsworth, though in obvious pain from his foot, insisted on walking back to the ship rather than allowing the others to tow him on the toboggan. When at last they came to drop down from the Barrier to the sea ice, they found the tidal crack had widened. Whales made a wonderful sight as they rose and breached in the water in the widened space between the blocks of ice, but it took the men even longer than before to cross the gap. The planks they had used in the morning to bridge it had floated away. One man nearly fell into the pool below as he tried to retrieve them. He was saved from the icy water just in time, and the others made a precarious bridge over which to tow the toboggan. Ellsworth followed behind them, his limp now quite pronounced.[5] The group eventually came aboard *Discovery II* in the late evening of Thursday, 16 January, little more than 24 hours after the ship had first arrived in the Bay of Whales.

Less than an hour before midnight on 16 January 1936, the moment for which everyone had been waiting at last arrived. Before them on *Discovery II* stood a man of slight build and of less than average height, with blue eyes and a sun-darkened face. Lincoln Ellsworth seemed clearly relieved to be aboard and reciprocated their exuberant welcome.

Later that evening Ellsworth recounted his own version of the events that he and Kenyon had experienced since they set out on their flight over the continent. As his listeners now learned, their having to sleep in their tent alongside the *Polar Star* at about 6,000 feet altitude had been more difficult than perhaps Kenyon had made out. Added to this, though their final landing site had been around 16 miles from Little America, it had nevertheless taken them nine days of tramping to reach their destination.[6]

Chapter 26 'NOT "RESCUED"-"AIDED"'

It was fortunate for Ellsworth that his relief had not been delayed. Though he seemed comparatively well and remained stoical about his foot, it was far more seriously inflamed and damaged than he was prepared to admit. He had assumed that the pain was due to frostbite, but the reason was different. Once established at Little America, Ellsworth and Kenyon had made a daily trek, over six miles each way, to Ver-sur-Mer, the romantically named vantage point above the Bay of Whales where Richard Byrd had unloaded supplies for his Antarctic expeditions.[1] Byrd landed at this spot after his flight to the South Pole in 1929, and named it for the place in Normandy where he had landed after his flight across the Atlantic two years earlier.[2] It was here, when they were first searching for Little America, that Lincoln and Kenyon had realised they were close to the Bay of Whales and the Ross Sea. Following the coast southward they found the beginning of a line of orange streamers originally fixed by Byrd, trekked alongside them and eventually made their way up a slight slope to Little America itself. Back at Ver-sur-Mer, the American aviators subsequently established the orange tent they hoped would eventually attract the attention of Wilkins and his companions on *Wyatt Earp*.[3] Their daily trek backwards and forwards on the round trip from Little America to watch out for *Wyatt Earp* became an essential part of their daily routine as they tried to ensure they would meet Wilkins and *Wyatt Earp* at the first possible opportunity.

It had already taken them ten days after landing at Roosevelt Island to locate Little America, and the endless walking had taken its toll on Ellsworth. His foot had become blistered earlier in the expedition when moccasins he wore at their second camp shrank with the endless damp.[4] Now, rawhide webbing from the sole of one of his snowshoes had worn loose and chafed his foot incessantly as he and Kenyon trekked to and from the lookout point.[5] It was no surprise that the original blister had developed further, and the area had become infected.[6]

Ellsworth and Kenyon had not been expecting their own party, led by Wilkins, to arrive before late January at the earliest. Even so, once at Little America, Ellsworth found himself fretting more and more impatiently as he and Hollick Kenyon waited for his friend and expedition leader to arrive with *Wyatt Earp*. His impatience was probably fuelled by the discomfort from his foot and the fever which developed as a result. By 15 January, a month after they arrived, Ellsworth's foot was giving him considerable pain the full length of his leg. He was running a temperature and had, by his own admission, become confused. It was in this state of mind that the news that they had been located first reached him. Though at the time Ellsworth would not acknowledge how worrying the situation might be, the infection had by then clearly taken hold to an alarming degree, and his temperature had been running at 102°F. He attributed his rejection of immediate help to a high temperature and a feeling of delirium.[7]

Hill gave Ellsworth a stiff whisky on his arrival, immediately after which John Strong, *Discovery II*'s doctor, attended to the explorer. Strong found that the blister covered much of the sole of Ellsworth's foot and some of his toes. The foot was generally swollen, and some glands were inflamed. The doctor treated Ellsworth appropriately with regular draining of the blisters and with hot foments, and so managed to stave off further spread of infection. Even so, it took three days for Ellsworth to recover.[8] Considering his own admission in his book, *Beyond Horizons*, about his explorations over the years, he was fortunate to receive medical attention when he did.[9]

Later that evening, after the first round of treatment, Ellsworth recounted his own version of the events that he and Kenyon had experienced as they flew over Antarctica. As his listeners now learned, the flight and the enforced landings had been considerably more difficult than Kenyon had made out. One factor was the intense cold they encountered at the higher altitudes. Starting their radio in the cold at 6,000 feet, let alone getting *Polar Star* airborne, provided serious problems. Managing to get sleep was also challenging, as they lay in the small tent they had pitched alongside their plane. Then their fuel ran out within sight of the Ross Sea and, though their final landing site was around only 16 miles from Little America, it had taken them nine days of searching and hard tramping through snow to reach their destination. Once at Little America, they began their long wait for the relief group to arrive, which they knew would not be before mid-January at the earliest. The date when Wilkins finally arrived was within the planned timescale for their expedition, but time had still weighed heavily on the two aviators as they waited.

Despite his obvious relief at the time at the arrival of the British-Australian expedition, Ellsworth was later at pains to state very clearly that *Discovery II*'s mission, in a steel ship no less, with all its attendant risks, had been unnecessary. All the plans he and Wilkins had made before he and Kenyon took to the air had been prepared with meticulous care. He had allowed his pilot and himself five weeks to cross the Antarctic continent — they had made the journey in three. Wilkins aimed to reach the Bay of Whales on or after 22 January 1936. *Wyatt Earp* arrived at their destination two days before the date they had anticipated. Ellsworth and Kenyon would only have had to wait another four days after *Discovery II*'s arrival. As Ellsworth put it in *Beyond Horizons*, he and Kenyon were not 'rescued', merely 'aided'. How much the scientists on *Discovery II* who had lost their planned work for the season, or the sailors who had risked so much in pushing through the Ross Sea ice, knew at the time of Ellsworth's own disappointment is not known.

Ellsworth and Kenyon's main problem was the failure of their wireless communications, the result of a faulty transmitter switch. Otherwise, it was clearly the result of good planning and navigation by air and land that they managed to spend the extra weeks in the dugout at Little America. The timing of Wilkins's arrival was in keeping with their plans and close to the intended schedule. Had the northerly trend of the ice edge and a strong wind not forced *Wyatt Earp* towards the north and away from the continent, his own ship might have reached her destination at the Bay of Whales sooner.[10]

In spite, however, of the careful planning and the nearly successful estimate time of arrival, the delay between Ellsworth's and Kenyon's arrival at Little America and Wilkins' arrival with *Wyatt Earp* could have led to an even more serious problem. As Hill wrote of Ellsworth:

> He seemed comparatively well but it was perhaps fortunate that his relief was not delayed. His foot was poisoned and though not very bad when we met him it might have had serious consequences within a short time. He received medical attention for several days.[11]

Timely treatment to Ellsworth's foot on board *Discovery II* may well have saved Ellsworth's foot from lasting damage, and his general health from serious decline. The 'aid' Ellsworth received may well have been more helpful than he either realised or was prepared to admit.

Chapter 27 AFTER THE RESCUE

In various parts of the world people were clamouring for any new crumb of information about the Antarctic rescue project. Hill telegraphed immediately to let the Discovery Committee know that Ellsworth and Kenyon were alive and safe. Secretary Borley had sent careful instructions about the way news should be spread round the world. As instructed, the first telegram was written in government code. Once they received acknowledgment from London that they had received word of the rescue, Hill was then free to let Wilkins, on *Wyatt Earp* know, after which Ellsworth and Kenyon could send brief personal messages. Hill had already been told by the Committee that this would be the time to send a full, official report of proceedings. After that, Ellsworth could send his own press reports.[1]

When the Discovery Committee sent this telegram, the ship had barely experienced the worst of the Ross Sea pack ice. There had been more than enough for the officers on board the research ship to consider. Now the telegraph room became busier than ever as each of them sent their reports in due order, as first queries, then congratulations flowed in.

In London, the *Evening Standard* of Thursday, 16 January was the first to publish the story. With the banners 'Lone Explorer seen from the plane may be Ellsworth' and with a large spread below, the evening newspaper announced to London and anyone who picked up the newspaper that a radio message had been received from a British rescue vessel. It spoke of an 'American Airman Missing Two Months' but remained cautious, going on to refer to a 'corrupt group' in the text of the telegram, words which could not be deciphered.

Next day, *The Times* of 17 January 1936, brought the news to the nation as a whole.[2] This, like the piece in the *Evening Standard* of the previous evening, was based on no more information than the Colonial Office had given the evening before, so the gaps in the story remained tantalising. Importantly, *The Times* reported that *Discovery II* had arrived at the Bay of Whales at 20.00 two days before and that the Australian airmen had, from their plane, sighted a man at Little America. A machine lay in the snow nearby. Like the *Evening Standard*, *The Times* wrote of difficulty in decoding some of the text of the telegram: it was impossible to tell whether the man seen on the ice was one of the two explorers for whom *Discovery II* was searching. *Discovery II* was now awaiting the arrival of Ellsworth's own base ship, *Wyatt Earp,* which was still making for the Bay of Whales. At that time *Wyatt Earp* was around 420 miles from the Bay of Whales and expected to arrive between 22 and 25 January.

There was a resumé of the news of the past six weeks. *The Times* recalled how, from 24 November, no news had been received from Ellsworth and Kenyon, then outlined both the reason and the route the American explorers had taken. It referred also to *Discovery II*'s departure from Dunedin with aeroplanes and sledges on board.

It was frustrating for the Discovery Committee not to have the full story straight away and another example of the problems of communication over 11,000 miles. However, from then on, more complete news began to reach them. Gradually they could piece together a fuller story.

The next day, 18 January, *The Times* carried a long article, which ran next to news of the King's illness and the death of Rudyard Kipling. Headed 'Rescues in Antarctic. Missing airmen safe. Fine work by *Discovery II*', it reported the news from further wireless messages received in London the previous day from the research ship. The two explorers, Mr Lincoln Ellsworth and Mr Herbert Hollick-Kenyon, Ellsworth's colleague and pilot, had been found fit and in good health after two months on the ice barrier near the Bay of Whales. The article gave the cause of their forced landing and explained why they had been unable to communicate with the rest of the world.[3] More followed on the next page, with a summary of the news of the explorers' last two months and of the international cooperation that led to the dispatch of *Discovery II* to find them. Other newspapers followed suit, both national and local, each with their own slant, especially where a member of the personnel on board *Discovery II* came from the immediate area of the publication.

From then on messages of congratulations and further enquiries flooded into the telegraph room in *Discovery II*. Over the previous weeks Hill had already found it necessary to send and receive far more wireless telegraphs than usual. Before *Discovery II* left Melbourne, the Commonwealth Government had asked Hill to let them know the ship's daily position. The Discovery Committee agreed that, in addition to this, the Commonwealth Government should be sent twice-weekly messages suitable for publication. Hill was to send the same to the New Zealand Government. The three-way communication between Wilkins on *Wyatt Earp* also took time.

As a result, A.E. Morris, the telegraph officer, had been working long hours, by day and by night, on the seemingly endless cables which came and went to and from the ship throughout the voyage. Now many more came from individual newspapers, each asking for their own details of the rescue. The replies to these were often brief, with an explanation that wireless difficulties prevented him from sending any long message and a referral to the Discovery Committee for fuller information.[4] The telegrams of congratulations which poured in required a more personalised reply. As a result, Morris was almost overwhelmed by the fresh influx of messages. Hill asked Andrew Marr, the scientist, to help him, with further support from Lieutenant Marchesi. These two readily agreed to provide support Morris to deal with the extra work which remained a full-time task throughout the days to come.

....

The scientists had immediately taken advantage of the delay as they waited for the shore party to return with Lincoln Ellsworth and had started to create a programme of their own. They had been deeply frustrated when they abandoned their work at the ice edge to the south of the Indian Ocean and realised that they would be unable to carry out their meticulously planned programme for this part of the commission. As the Discovery Committee had realised from the moment the idea was first mooted that a rescue should be undertaken one of by their own research ships, the scientists were now seriously thwarted in their research. They would be unable, during the period of the rescue, to collect the material for which they hoped, and as a result would lose the continuity crucial for them to analyse their findings. In addition, they were now in a completely different part of the Southern Ocean from the area first intended.

When first summoned back to Melbourne, Deacon had, rightly in the circumstances, said he would relinquish all stations until instructed otherwise.[5] However, his disappointment was immense. To compensate for this, he had, not surprisingly, hoped to make good in other directions. Even on 28 December, Deacon had asked yet again for permission at least to make a line of full stations between Dunedin and the Bay of Whales. He stressed the importance of both hydrological and biological observations at the time of year they were making the voyage. This might, Deacon had added, be easily feasible if the line the Discovery Committee had initially requested, following a great circle, were altered and they postponed their report of the position of Emerald Isle until another time.[6]

Both for the sake of conserving *Discovery II*'s fuel supplies and to allow for a direct route to Dunedin from Melbourne, the Committee had been prepared to forgo the passage that they had first considered, past the supposed Emerald Isle. However, they refused Deacon's request to make a line of stations from Dunedin to the Bay of Whales. As late as 30 December, the Committee confirmed that the chief scientist and his colleagues might well find opportunities for making stations if held in the ice for any length of time or delayed at the Bay of Whales. Then, and only then, might they undertake scientific work. They must not consider it before that. They would be able to take soundings on the return journey.[7]

It was understandable that the scientists were disappointed. There had been one compensation on their voyage south from Dunedin. In the pack ice the scientists had enjoyed a form of freedom to which they were unaccustomed. Liberated for the time being from both the microscopes in their laboratories on board ship, and the bureaucrats who oversaw almost their every move, they could spend the day on deck, watching events as they unfolded in the ice. They had even enjoyed the time when they formed part of the poling party. Exhausting work though that was, it made a complete break from their normal work. Although the Committee had suggested they might take soundings if the ship were delayed for long in ice, there was certainly no time for that. The poling work was essential if they were to protect the rudder, and ultimately proved rewarding. But, whatever their enjoyment at that time, they were excited and relieved when they finally received permission to resume research, and eagerly took what advantage they could of the situation now that they had reached the Bay of Whales.[8]

From the moment they first arrived in the inlet, they were amazed to see the number of whales and orca all around the ship. The Bay of Whales was still living up to the name Ernest Shackleton had given it when he was on *Nimrod* in 1908. Minke whales were diving on all sides of the ship, scooping up the krill everywhere, before returning rapidly to the ice edge. Orca, known as killer whales, sported in the Bay, as well as along the edge of the Barrier.[9] Clearly there was an abundant supply of krill. The scientists' minds raced to the research they might do, even as they concentrated on the search for Ellsworth.

Their first move, as they waited in the Bay for the shore party to come back from Little America with Ellsworth, was to conduct a full station within the Bay, sifting the collection of copepods and other fauna they hauled up. They soon realised how extensive the swarms of krill were. When their nets came up, they were completely choked with mud from what was apparently a dense deposit on the bottom of the bay. Thick sludge and sediment soon covered the deck of the research ship. As the seamen began to sluice down the deck to clear the mud, vast quantities of krill poured out from the ship's hose and covered the deck.[10]

For everyone's convenience, the scientists quickly decided to haul the nets into a boat and sift the contents from there.[11] The air temperature was around 20 degrees Fahrenheit, the sea temperature just on freezing, but the scientists seemed unperturbed by the numbing cold as they sat and worked in their small boat. Back in their laboratories they soon realised there was great scope for more detailed study.

Discovery II lay to off the Bay of Whales for another night. The next day, Friday 17th, with Ellsworth now on board and recovering in his cabin, there was spare time to occupy until *Wyatt Earp* arrived. Hill and Deacon decided to make use of the time and to carry out a fuller scientific survey. At 07.30 they set off from the Bay of Whales on a north-westerly course. Throughout the afternoon the scientists completed several stations in the Ross Sea. With overcast sky, a moderate breeze and air temperature now around 22 degrees Fahrenheit it was a chilly return to the long, cold hours they normally took as routine. Around 18.00, frequent snow showers began to pass by. They could see little more than half a mile ahead, and the ship remained not far from the Barrier, the scientists working regular stations throughout the night and well into the next morning.[12]

On Saturday, 18 January, with little hope that Wilkins would arrive in *Wyatt Earp* that day, they remained at sea, steaming slowly along, about two miles out from the face of the Barrier. The day started well, the weather far better than on the previous day. The sun appeared from time to time through broken clouds, and the ship rolled slightly in the swell. Later in the day, with the wind blowing from the south-east, the temperature dropped to 14 degrees Fahrenheit. As the cold air passed over the sea, it became laden with ice crystals and, while the sky above remained almost cloudless, what seemed like fog or white smoke, the phenomenon known as sea or frost smoke, began to form. Caused by cold air moving over slightly warmer water, it rose to about three feet, making it difficult to see the water surface ahead. Hill decided to stand out to sea and wait for better conditions. Soon, in view of the forecast that these conditions would continue, he planned to postpone their return to the Bay of Whales until Sunday. It was a wise move, as the frost smoke remained throughout that night and well on into the next day.

On Sunday, a fresh breeze still blew from the southeast and the weather remained disappointing. The temperature only rose from a low of 13 to 18 degrees Fahrenheit at midday, before dropping back to 12 to 16 degrees Fahrenheit. The air was still laden with ice crystals, and the heavy frost smoke continued to fill the air.

The poor visibility was frustrating for navigation, but for the scientists it was an ideal time to start to analyse their findings. For the airmen, it was a difficult time, as they had little to occupy themselves while they waited for *Wyatt Earp* to arrive. They had expected to have much more to do once they reached the Antarctic. Now, with the aim of their part in the voyage achieved so quickly and further flying not required, they may well have felt deflated. Douglas had made an outstanding flight in difficult circumstances but no doubt was disappointed that he and Murdoch never had the chance to take off in the Wapiti, however difficult the conditions might have become.

Douglas was, however, soon cheered by an enthusiastic message of congratulations from the Prime Minister of Australia. He had already received a warm message from Sir Douglas Mawson, with congratulations for himself and his team as well as admiration for the Americans' magnificent flight. He gave Mawson's telegram to Ellsworth for him to keep. Douglas found little interest in the land around them. With little else to do, he was buoyed by the prospect of going overland to Little America and visiting the American base on foot, He also looked forward to the possibility of helping with the rescue of the Ellsworth's Northrop Gamma, *Polar Star*. Apart from these activities, the most interesting part of his job had been achieved. He was anxious to start for home as soon as possible.[13]

Chapter 28 Wyatt Earp

The frost smoke lingered for most of the day as *Discovery II* cruised along the edge of the great ice shelf. In the afternoon, the clouds cleared for a short moment. The Barrier was bathed in beautiful sunlight which to some extent compensated for the cold they all felt. The air temperature remained above 12 degrees Fahrenheit, but the wind chill made it seem much lower. Everyone, British, Australians and Americans alike, now wanted to know the next steps, wondering when they would leave the Bay of Whales, to which port they would return, and on which ship the Americans would return. Only when *Wyatt Earp* arrived would they have answers to all these questions. Ellsworth's ship seemed to be a long time in reaching them. They hoped, for the sake of Wilkins and his men as well as their own, that there had been no great mishap on their journey to cause *Wyatt Earp*'s people a long-term delay in the pack ice.

Ellsworth would not only be delighted to see Wilkins at last, but no doubt had other reasons for wanting to see *Wyatt Earp* in the Bay of Whales. Intermittently over the last three years she had been his main base. The one-time Norwegian sealer he had bought three years before and reinforced with strong metal sheeting, had already resisted considerable pressure from ice floes and pack ice. Moreover, she could cruise for around 9,000 miles. She had already proved of good service to him and his companion, Wilkins, on their two previous expeditions to the Antarctic. She was a sturdy vessel on whom Ellsworth felt he could rely.[1] Named for his hero, the frontier marshal and sheriff of that name, *Wyatt Earp* had become a much-loved home.[2]

Wyatt Earp had some limitations, one of them most noticeable in late 1935 and January 1936: she could not deliver the speed necessary for a rapid rescue. Her maximum speed was between

7-8 knots.[3] Her passage to the Bay of Whales from Deception Island off the west coast of Antarctica, first working her way over the Bellingshausen Sea and then moving along the coast of Marie Byrd Land, would not have been easy. In addition to her limited speed, all the waters through which she passed were heavily infested with ice, which slowed her even more. Weighing 400 tons, she was probably too light and her bow too blunt to push through all but the lightest of the floes which had massed together along much of her route. Ellsworth later attributed the delay to Wilkins' fear of over-working the engine. Though she arrived at the Bay of Whales at very much the time Ellsworth had expected, these drawbacks, and the inevitable delays she encountered on her way, undoubtedly contributed to her arriving at their rendezvous later than *Discovery II.* Perceived delay or not, Ellsworth would be relieved to see the ship which was, after all, the base for his expedition.

Patience was eventually rewarded for those waiting on *Discovery II.* In the afternoon of Sunday, 19 January, the clouds cleared at last, and the barrier was again bathed in sunlight. Then, just before 18.00, they spotted a small ship steaming eastward, *Wyatt Earp* herself. *Discovery II* made her way towards the vessel and, around half an hour after the research ship had first sighted her, the two ships were within half a mile of each other. The research ship's motorboat went across and quickly brought Sir Hubert Wilkins aboard. The ships made the remaining miles to the edge of the Bay of Whales in convoy, *Discovery II* just astern of *Wyatt Earp.* Wilkins then spent the night on the already crowded research ship.

Wyatt Earp seemed to some who now saw her as somewhat round and rather small, an unlikely vessel to have made the journey of over 2,000 miles from Deception Island to the Bay of Whales. The diesel engine was set well aft to make way for a long well deck in which they carried their own seaplane, another Northrop Gamma and companion to Ellsworth's own.[4] The float gear for *Polar Star* was still in position, left there when Ellsworth and Kenyon set out on their long flight from Deception Island.[5] There was little room to spare.

Wilkins, Australian by birth, had a reputation for derring-do and dashing exploits. In spite of his background as a war photographer, Arctic explorer and pioneering airman, which gave him an almost inevitable air of bravado, he gave the impression of being a pleasant and ebullient, indeed a charming person. Those who sat beside him over tea soon felt drawn towards him. It seemed that Wilkins was irritated because *Discovery II* had arrived before him at the Bay of Whales. At first, some of those listening to him understood his disappointment. When it later emerged that Wilkins himself had asked for help, the listeners were less sympathetic. As Douglas wrote, 'He asked them! He could hardly blame the Discovery Committee and the Australian Government wanting to carry it out properly or not at all.'[6]

On board ship together, Hill was relieved to find that he and Wilkins soon developed a cordial relationship. When they were both at sea, communicating only by telegram, Wilkins had, at least on one occasion, asked Hill to pass on information he could then send to the North American newspapers. The North American Newspaper Alliance and the New York Times had exclusive rights to accounts of their activities, and Wilkins felt sure that they would be pleased to pay for the information. If Hill were to send him exclusive accounts, he would then send them on to the newspapers linked to them. Hill, however, could not pass anything of interest to Wilkins without the Discovery Committee's approval and agreement. He suggested that Wilkins should contact the Committee directly. Later, at a point when *Discovery II* was at risk of becoming ice bound, Wilkins' enquiries had frequently reached *Discovery II* when Hill was too busy to reply. When he did reply, he said that, as and when they found Ellsworth, Wilkins would be able to send press messages as soon as Hill had been in touch with the Discovery Committee. Wilkins had replied that he fully understood the restraints that would be set by the Discovery Committee, but Hill had naturally remained anxious that this ruling would cause no tension between them.

The three-way correspondence had added one more demand on *Discovery II*'s time when the load of telegraph messages to and from the research ship was already heavy. Fortunately, such telegrams had generally moved quickly through the system. The Discovery Committee replied promptly whenever possible, and Hill was able to ease a potentially difficult situation. Ultimately, the two ship Masters reached a good understanding.

Most important now were the decisions to be made for the return journeys of the two ships. It was soon agreed that Ellsworth would return to Australia on *Discovery II*, as he was keen to express his appreciation to the Australian government for their efforts on his behalf. Only then would he leave the research ship and sail for America. Kenyon, meanwhile, would transfer to *Wyatt Earp* and stay for the time being in or near the Bay of Whales to help in the salvage of the abandoned *Polar Star*. Once that was completed, Wilkins would make straight for America, taking Kenyon with him.

Discussions lasted well into the night. Having moved a little way out to sea, the vessels returned in convoy to the Bay of Whales the next morning, in the early hours of 20 January, at 04.00. *Wyatt Earp* lay tied up to the Barrier ice while the men on board *Discovery II* conducted navigation tests and practised their own lifeboat drill. The wind still blew from the south, but it had dropped considerably, and the ships then lay a short way from the ice, in the lee of the cliffs. The temperature rose to 28 degrees Fahrenheit, making 20 January the warmest and brightest for several days.

Several men went ashore in the whaler and walked on skis or snowshoes the four and a half miles to Little America. They found Byrd's settlement situated in a small dip which protected it well from the winds which blew constantly during the winter season. On arrival the men went below, explored the quarters and saw for themselves the small township, buried under snow, with only the wireless tripods and masts appearing above it. Initially, when they first reached this goal. the American explorers had found it supremely welcoming, a crucial place of refuge. Later it had come to seem confining and constricting. To some of the people from *Discovery II*, it appeared a dark and gloomy underground retreat, sad and lonely now that the work of the Byrd expeditions was over.[7] They could understand why Ellsworth and Kenyon had been eager to leave. To others from *Discovery II*, it seemed like a small city.

Much of the food that Douglas and Murdoch had dropped from their Moth five days earlier still remained uneaten. Hungry after their early start, the visitors devoured it, then returned to the Bay of Whales and the two ships waiting there.[8] With the sun high in the sky and with little wind, the return trek left them unusually hot. At 18.00 they went to *Wyatt Earp*, where they received a friendly welcome from Wilkins. Once on board, they were again struck by the small space the ship's design allowed them. In spite of the cramped space, they enjoyed a pleasant evening. *Discovery II* lay to off the Bay of Whales, where they returned in the motorboat at about 23.00.[9] That evening they heard of the death of the British king, George V, and the flags on both ships were lowered to half-mast.[10]

....

Hill decided not to start the return journey to Melbourne too soon. He may well have hoped that he and Walker could catch up a little more on the sleep they had missed on the southward journey. He also wanted space before their next encounter with pack ice, to make careful plans for the route they would take for the return journey. Deacon, too, was ready to stay. The hydrology and biology of the area was markedly different from the rest of the Southern Ocean. It made sense, now that they were in the area, to conduct as thorough a survey of the southern part of the Ross Sea as they could. When the Scientific Subcommittee in London first recognised that the circumpolar cruise intended for the fourth commission was to be abandoned, they had immediately recommended that, when the search for the American airmen was completed, the ship would work westward across the Indian Ocean towards South Africa.[11] The Discovery Committee accepted that they could return from the western side of the Ross Sea. With this in mind, Hill and Deacon made plans for the scientists to make a series of stations as they worked their way westwards, before eventually turning north.

They stayed in the Bay for one more day, giving those who so wanted another chance to visit Little America. The scientists carried out another station. The weather was considerably more pleasant than on the first day when they had spent time sorting krill in the Bay of Whales. The temperature rose briefly above freezing point, there was little wind, and the sun made an appearance from time to time.

That evening, the officers and scientists in *Discovery II* gave a final dinner for Ellsworth, Kenyon and Wilkins, in the research ship's wardroom. Conversation ranged widely until soon after midnight when the last farewells were said. Ellsworth, though he was unwilling to admit it, was still not fully recovered, and settled to stay on the research ship for the journey back to Melbourne. Wilkins and Kenyon would salvage the *Polar Star* before leaving the Bay of Whales, and then prepare to return in due course on *Wyatt Earp* to the United States.

The people on *Discovery II* were disappointed that Kenyon had to leave their ship. Even in the short time the Canadian had been with them, he had impressed everyone with his modesty, his cheerful nature and his ability to settle into their routine so well. He was well-liked for his relaxed manner and dry humour. Kenyon himself was disappointed not to be able to stay on *Discovery II*, but fully realised the need to find *Polar Star*. It also occurred to him that he might well reach America sooner than his fellow explorer.

In due course, when *Discovery II* was well into her return journey to Melbourne, Wilkins sent news to *Discovery II* that, along with the aviator on board *Wyatt Earp* and two others from the ship, Kenyon had managed to locate *Polar Star*.[12] The four men dug the plane out of the snow and Kenyon flew it to the Bay of Whales from where it returned home on board *Wyatt Earp*. Kenyon's move to the American ship had been worthwhile but those who returned to Melbourne on their own ship would miss his balanced judgment and sense of proportion.

Chapter 29 Return through the Pack

Discovery II set sail from the Bay of Whales in the early hours of Wednesday, 22 January. Soon the sun was shining through broken clouds. The ship rolled slightly in the swell. It was the start of a number of pleasant days in the Ross Sea as they worked their way towards McMurdo Sound. The ship crossed the 180th meridian, the date line, on Thursday 23rd. At midday on Saturday, 25 January time was adjusted to noon.

Over a period of four days, the scientists completed two stations each day, making a satisfactory number of lines in all. These findings made a helpful addition to the work they had already done in the Bay of Whales and contributed substantially to their research on the absence of a warm deep current in the Ross Sea. The weather remained good all the time they were there, but these stations brought a different set of challenges from those they would normally experience. The scientists had surrendered much of the storage space usually reserved for them in the holds and had put their heavy equipment ashore in Williamstown to make room for items required for the rescue mission. Now aeroplane parts filled areas below decks where normally spare nets were stored. Above, extra deck cargo had replaced the haulage gear men usually used to hoist back on board the bottles they lowered when collecting samples. Without the Samson post in its usual place, it now took much longer for the scientists to haul them in, and they experienced some serious damage to them as a result. The samples that survived had to be piled high in the laboratories. Here too, goods for the expedition had filled spare corners, and the scientists were now limited by the cramped space left to them.[1]

Despite the difficulties, the research proved worthwhile. Between them, Deacon and his colleagues were able to establish several important points. Echo sounding confirmed a shallow submarine ridge which ran from Cape Adare to King Edward VIIth Land. This retained the cold water in the Ross Sea and prevented the warm, deep current from the Southern Ocean from entering its most southerly part. It could well explain why several species were missing in the South, especially the krill, where a different species, *Euphausia superba,* found in the ocean to the north, was replaced by the *Euphausia crystallorophes* which they had first seen in the Bay of Whales. This finding might also, they felt, highlight the importance of the deep current which flowed around the entire Antarctic continent.[2] They hoped to do more research in that area in the future but knew this would not be for some time, if at all.

On 25 January, they completed their fourth line of stations off Ross Island in the far west of the Ross Sea. The ship approached Cape Crozier, at 77° 31′ S 69° 23′ E, the island's most easterly point, soon after midnight. The basalt columns of the cliffs below the slopes of Mount Terror stood out, etched in the midnight light. Here the scientists landed to collect some rock specimens and returned to the ship at 02.00. After this, Deacon was keen to examine the conditions in McMurdo Sound where most of the earlier biologists had made their collections, and then work in one of the deep basins near the coast of Victoria Land.[3] To enable this, Hill considered taking a course to pass through McMurdo Sound, and then pass Cape Bird, the furthest north west extremity of Ross Island, at 77° 13′ 05″ S 166° 26′ 09″ E, after that Beaufort Island, and finally Franklin Island.

But as he considered working their way in that direction, Hill soon was compelled to abandon any plan to spend time there. They were already in challenging waters. This was an area full of history. In 1841, Ross had named the different features of the archipelago for members of his crew, for his ships HMS *Erebus* and HMS *Terror* and for others connected with his expedition. More recently, in 1911, Scott had fixed the base for what would be his last expedition at the tip of the southernmost promontory. It would have been satisfying to linger in the area while the scientists conducted their research. However, Hill was well aware that the first *Discovery*, on Scott's initial expedition, had been locked for two years in the ice off what became known as Hut Point. He had to admit that he could well be confronted by ice on the southern side of Ross Island well before *Discovery II* ever reached this area. Rather than take this risk, Hill turned north after leaving Cape Crozier, then proceeded westward, sailing about ten miles from land, along the north coast of Ross Island. He and Deacon still hoped at least to reach Cape Bird, the island's most north-westerly point, and explore McMurdo Sound from the north.

Once more, they were thwarted by the pack ice they found ahead. The sea beyond Cape Bird and Beaufort Island was closely packed with ice. There seemed no limit to it, and it seemed reasonable to assume that they had reached the approaches to the southern corner of the Ross Sea pack. Hill turned to the north and headed towards Franklin Island. They viewed the island from as close as possible but again pack ice hindered their movements. Now the ice seemed to stretch well towards the coast of Victoria Land. The scientists had hoped to move on to survey at least one of the deep basins near the coast of South Victoria Land, but, here too, soon abandoned plans for research. It was one of the areas which Hill knew to avoid, where pack ice would accumulate and drift slowly northwards. The spectres of the *Southern Cross*, *Terra Nova* and *Aurora* once more rose to the forefront of his mind, with their warning of ships which, once locked in the pack, then drifted for weeks or months, unable to escape its grip.

Coupled with the impending problems of another journey through pack ice, fuel supplies were now dwindling. This was clearly the moment to bring the current run of stations to an end, and Deacon abandoned thoughts of any further observations in that area. Hill knew well to move from this area as soon as possible. With the summit of Mount Terror on Ross Island still visible almost 70 miles away, he turned northwards, and *Discovery II* began her return journey to Melbourne.[4]

It was disappointing for the research staff, but for Douglas and Ellsworth, at least, it meant the start of their homeward journey. They, and no doubt the other Australian airmen, had not appreciated how long-drawn-out their voyage on the research ship would be. Between themselves, Douglas and Ellsworth agreed it would be encouraging to move on without the frequent stops they had been making so far on their return voyage.[5] The stations and the streaming of the speed log would continue, but at least they were finally heading for Melbourne. The ship reached the southern edge of the 500 miles of pack ice which guarded the Ross Sea from the Southern Ocean at midnight on January 27th, at 73° 37′ S 176° 43′ E.

Hill surveyed the seemingly endless ice to their north with a mixture of feelings which ranged from deep foreboding to sheer determination to make their way through. At this point they were navigating through loose pack ice, nosing their way gently through floes, opening up small leads between the floes, and experiencing frequent, now familiar clashes with ice. They returned to the familiar feeling of suspense as they waited to know whether these clashes had led to any serious damage. Their sight was sharpened to check out what leads might lie ahead of them.

Douglas and Murdoch welcomed their chance to take to the air once more. The airmen made two flights in the Moth to inspect the pack ice on the morning of 28 January. Already familiar with the procedure for getting into the air, they taxied round in a pool of water to warm the engine, then took off with relative ease. As they circled up and flew around, they saw a stretch of open water about 15 miles away. *Discovery II* reached the spot early that afternoon but a few hours later met more pack ice. It was still possible to draw on the morning's aerial reconnaissance, and Hill took what seemed like a clear route to the north-east. Their movement forward remained, however, slow going, and that evening Douglas and Murdoch made another reconnaissance flight in the Moth. By viewing what lay ahead of *Discovery II* and being able to report on the distance and area of ice they could see, the Australian airmen again proved their value beyond measure.[6]

Douglas learned and recorded much from these flights. As he and Hill would claim later, a more powerful engine might have made take-off easier. A ship's derrick might have helped, provided it was big enough, to swing the plane further out with engine already running. Already, however, they had confirmed the already growing awareness that the future of reconnaissance in the Antarctic lay in a combination of search and navigation by both ship and by plane.[7]

The journey on *Discovery II*'s return through the pack ice was steady and generally predictable, apart from one period in the early morning of 29 January. At that point the ship encountered larger floes, easily comparable in size with those they had seen on the southward voyage. At first sight, these floes caused some concern, but the openings between them were wider than they had seen between the massed blocks of ice they had observed three weeks before, and this time the floes brought less fear. Once *Discovery II* had passed this area the ice opened up, the leads spread out and the floes became much smaller. But Hill and his colleagues knew not to be complacent, and wisely so. Almost inevitably, as evening approached, heavy ice slowed them once more. It was around 15.00 the day later, 30 January, at 71° 57′ S 178° 38′ E, that Hill recognised a welcome sensation as the ship began to move gently but distinctly in response to a soft, northerly swell, and he realised that open sea was not far away. His heart lifted, but he knew their escape from the ice would not be immediate: it still took them some time to work their way through the last of the heavy pack.

They reached open water at around 18.00 in 70° 5′ S 178° 40′ E. This was a mere three days after they first encountered the pack on their return journey. They had made the southward journey in unexpectedly good time. Their return, taking two days less than that, was remarkable. It may well have been the shortest journey through Ross Sea pack ever taken.[8] *Discovery II* remained hove to at 18.00 for the scientists to make another station. A heavy snowstorm in the late evening blinded them as they came to the end of the station, but even this did nothing to dull their sense of liberation from the heavy ice. At midnight, with the wind blowing from the north-west, *Discovery II* was once again pitching easily in a gentle swell. With regular stations and frequent streaming of the speed log, they their normal routine was resumed. Their relief was unmistakeable as they considered how much their steel ship had survived.

Chapter 30 Winding Down

Hill called to steer northwards for a day. Then what seemed a good opportunity presented itself and he turned westwards in the hopes of surveying the Balleny Islands which lay around 150 miles to the north-west of Cape Adare. The Balleny Islands consisted of three large islands and several smaller ones which formed a chain running in a generally NW-SE direction for about 100 miles. Discovered by John Balleny in 1839, the original charts were still used as the standard reference for navigation in those parts. Of volcanic origin, like Scott Island to the east, these islands were frequently beset by pack ice from the Antarctic continent. Encouragingly, on this occasion the islands were free of surrounding pack ice. Hill, who had long wanted to study the area, initiated some survey work.

It proved intriguing. So much of what Hill and his colleagues saw differed from Balleny's report in 1839 that they soon realised that they could not confirm those first observations: the islands appeared much closer together than charted, while the 12,000-foot peak marked on Young Island did not exist. Young Island and Buckle Island rose gently to about 4,000 feet. There were no peaks, nor was there any recent sign of volcanic activity.[1]

Other work was unfortunately broken short. They launched the motorboat, and a party of scientists left the ship to make a landing on Borradaile Island, a small island a little way off the most northerly island. It had a spit suitable for landing. There they planned to collect rock specimens. They had only been away from the ship for about half an hour when heavy snow began to fall, blocking their visibility. *Discovery II* was equally affected. Hill immediately recalled the party in the motorboat, guiding them back by means of the ship's siren.

This was a worrying time. Not only was visibility very poor, but alarmingly the motorboat failed to start. The engine had never been totally reliable and had frequently caused problems. It eventually emerged that it had not been properly overhauled during the maintenance of 1935. While the engine appeared to be in good condition, the ignition and the pump had clearly not been touched. The situation was critical. The pram dinghy was sent to tow them back, and, so heavy was the snow, blasts from *Discovery II*'s siren were essential to lead them in. The party arrived back on board at 16.47 without having made a landing. Hill then steamed clear of the islands and lay to, hoping for the sky to clear. The weather, however, remained thick, and at 04.32 on 5 February they abandoned the survey. It was disappointing to have to leave the Balleny Islands, but the risks had proved too great. The survey was a project to which Hill would return two years later, with considerably greater success.

After they left the Balleny Islands, they sailed northwards, taking stations at regular intervals, 100 miles apart. For scientists and marine staff alike, it was good to be able to keep to a steady pattern. Ellsworth expressed his pleasure at the interest of the journey and at the comfort afforded by the Chief Scientist's cabin, which he occupied for the journey.[2] He was, not surprisingly, disappointed, as Douglas records, that they were taking so long over the journey.[3] However, Ellsworth was pleased, as he put it, that 'it was not all wasted time for the research vessel'.[4]

For the airmen, the delays due to the time spent on the stations proved tedious.[5] They found time passed very slowly, though they soon found things to do. They had, at least, to attend to the Moth and the Wapiti, and from 8 February, during the final stage of the return journey, they were busy removing the aeroplanes' wings and checking spare parts.[6] Two of them, Flight Lieutenant Douglas and Sergeant Easterbrook, also spent time carefully making models of the Moth, which they presented to Ellsworth and Hill, tokens much appreciated by their recipients. They had already given Ellsworth the parachute the Australians had dropped at Little America, inscribed with compliments to

> MESSRS ELLSWORTH AND KENYON FROM THE PERSONNEL OF THE ROYAL AUSTRALIAN AIR FORCE ABOARD R.R.S. DISCOVERY II BAY OF WHALES ROSS SEA ANTARCTICA JAN 15TH 1936.[7]

This Ellsworth later presented to the National Museum in America, along with his model of the Gypsy Moth.[8] Hill's model resides with the National Oceanographic Library in Southampton.

Poor weather was another factor in slowing their journey and in delaying their return. Once north of McQuarrie Island they were soon reminded that they were now back in the Roaring Forties. The westerlies which circulate the southern hemisphere at these latitudes had treated them with reasonable kindness on their voyage south. Now they hit *Discovery II* with full force. From 8 February *Discovery II* battled with strong north-westerly winds, and progress was slowed considerably.[9] Fortunately they passed two miles east of the formerly charted position of Emerald Island before the storm reached full force. The Discovery Committee had been anxious for them to identify it and confirm its position. Visibility at the time was around ten miles. It was easy to take the bearings and make soundings to establish the depth of water as they passed the supposed island, but they could not confirm its existence.[10]

By early evening on 11 February, a very high sea was running. From 18.53 they lay hove to in gale force winds, the ship pitching into the deep trough which followed each wave, then returning seemingly skyward. It meant a rough time for everyone on board, and it was not until after 09.00 the next morning that conditions began to ease. By midday on 13 February the wind had changed to a moderate south-westerly, but not before their planned return to Melbourne was delayed by nearly two days. Again, as on the outward journey in the Bass Strait, the stowage of aircraft and stores proved sound, though the wild sea sent an unexpected amount of spray over the Wapiti.

Ten days after leaving the pack they passed along the east coast of Tasmania. It was high summer. Bush fires raged on land, and the smell of woodsmoke wafted towards them, giving everyone, not only Ellsworth and the Australian airmen, a great longing for the shore.[11] A cheer went up when in the distance they saw the great, granite lighthouse at Eddystone Point on the north-east of Tasmania. It was now 15 February, the time 15.53. Albeit a mere speck to begin with, it grew in their eyes with comforting rapidity the nearer they got. They went through the Banks Strait leaving Flinders Island to starboard and entered the Bass Strait in the late evening.

Even then challenges still remained. The wind freshened towards midnight and the ship pitched heavily for some hours as they worked their way forward into a strong south-westerly. The movement was uncomfortable enough, but this was a different journey from the one they had made eastward along this notorious stretch of water six weeks before. This time the Bass Strait held few concerns for them and they soon moved swiftly westward at around ten miles per hour. They instinctively accepted the moderate rolling. Even the pitching was acceptable. This was homecoming for everyone as they checked off the various landmarks of the Strait as they passed them.

That evening, scientists, Australian airmen and marine officers enjoyed a last dinner at sea. Then followed celebrations of a different kind. Deacon announced that the King had approved Hill's appointment as Officer of the Order of the British Empire (OBE).[12] Hill had been sounded some days before and the Commonwealth Governments informed.[13] The news was official and by then had been made public. Douglas rose to congratulate Hill on behalf of the Australian Air Force members present at the dinner. Hill returned the compliment, saying that he accepted the honour on behalf of everyone on board and thanked them all warmly for their enthusiasm and support. The Discovery Committee had moved very quickly to secure the appointment, a civilian OBE. Its announcement added to a cheerful end to a memorable journey.

From the start, Hill had found Douglas and his Australian colleagues an invaluable help in the enterprise. Not only had the voyage through notorious pack ice been transformed by their ability to see over many miles of generally impenetrable ice, and so recommend a reasonable route forward. Early on, Hill had also appreciated their 'no-nonsense' approach as being both helpful and encouraging. His own skill in navigation under difficult ice conditions had been immensely enhanced. Hill's response fully acknowledged this.

They left Cape Schank to starboard in mid-afternoon the next day, Sunday 16 February, and entered calmer waters. Just before 17.00, when they were a little way off Port Phillip Heads, they slowed down for the pilot to come aboard. He greeted them with a broad beam, pleased to be the first on board the now-distinguished research ship. He handed Hill a newspaper which gave a clue to the welcome *Discovery II* would receive when eventually they reached Melbourne.[14] He took them forward at full speed for a while, then guided them through the Rip and across Port Phillip Bay towards their destination.

Already they could see crowds cheering them on shore. Passing ships welcomed them in delight. One steamed alongside for several miles, the people on board cheering *Discovery II* all the way. It was a foretaste of the tumultuous reception which awaited them the next day. *Discovery II* anchored off Williamstown for the night. After the usual customs and medical checks, they tied up alongside Nelson Pier, the main pier in Williamstown, a minute before 10.00 on Monday 17 February, six weeks after leaving Dunedin for the South.

Chapter 31 MELBOURNE RE-VISITED

Before *Discovery II* had even reached the quayside a huge crowd of people surged forward to find a place as near to the ship as possible once she tied up. Nearby, a large group of schoolchildren waved small Union flags. A line of policemen tried to hold everyone back, but with little success. Ellsworth was persuaded to stand by the ship's rail, and cameras flashed as the press did their best to take shots of the distinguished explorer. The gangplank was lowered and marine staff and scientists, all smartly suited for the first time in over six weeks, followed the American to a dais set ready, close at hand. There, Ellsworth, closely followed by Hill, was welcomed by a reception committee. Sir Douglas Mawson was among them. Raymond Priestley, a one-time member of Scott's expedition of 1910–1913, and now Vice Chancellor of the University of Melbourne, accompanied them.[1] David Orme Masson, another keen supporter of Australian Antarctic exploration, was also in the group. They would all be among the first to visit the ship.

After the initial welcome, both Ellsworth and Hill were invited to make a broadcast to the world. Ellsworth spoke first, saying how pleased he was to be back in civilisation once more. He acknowledged the international spirit of goodwill which led to his return, then thanked the British Government, the Discovery Committee, and Lieutenant Hill and his ship, *Discovery II.* He recognised the part played by the Australian government and the men of the RAAF, the drone of whose plane was the first intimation of their release from their icebound home. He thanked the New Zealand government for their assistance in providing a base from where *Discovery II* could set out, and for the material goods and supplies they had provided.

Ellsworth added that, as a citizen of the United States of America, he hoped his success in making the trans-Antarctic flight might help people everywhere who had an interest in geography to widen their knowledge of the world in general and the Antarctic in particular.

Hill spoke next, saying that the success of their voyage to the Bay of Wales was reward enough for himself and his ship's company, and ample recompense for the anxiety, the work and the hardships they had endured over the last two months. He congratulated the Americans on a truly remarkable flight and thanked the Discovery Committee, the Government of the Commonwealth of Australia, and his own ship's company for all their assistance. He appreciated the minute detail that had gone into the whole exercise.

Hill then, typically, made light of the voyage, saying that, considering that it was a cruise in polar waters, their passage to the Bay of Wales had not been eventful. They had made a speedy journey through the ice. The sea beyond the pack was calm as they approached the Barrier and made for 73.05° S, just 420 miles from the Antarctic Circle. He referred briefly to the worry they experienced when on seeing the coloured streamers flying on the lonely spot above the Bay, on the surface of the Barrier: he had taken them as a signal of distress and feared that they might be too late to find the missing Americans alive. It was Australia's own airmen who had brought the welcome news that they had been able to communicate with the lost explorers.

Hill added that, having achieved the purpose of their voyage, they still had sufficient fuel to allow them to carry out some scientific work in their accustomed way. He was pleased to say that Mr Deacon, the Senior Scientific Officer, had assured him that as far as the main object of their work was concerned, the interruption of their original program, ambitious in scope as it was, would prove a blessing in disguise.

Hill ended by thanking everyone in Australia for the outstanding welcome they had received that day. He also thanked those who would hear his words broadcast in the UK for their kind messages. of support and congratulations. He ended his words of thanks by saying what a pleasure it had been to work with the Australian and New Zealand Governments.[2]

The speeches were broadcast to the English-speaking world, and for several months articles and illustrations of the expedition appeared in a variety of newspapers, journals, newsreels and films. They provided some encouragement to a world still concerned by the economic restraints in existence after the Depression of the early 1930s and already witnessing the rising threat from Nazi Germany.

Once the official welcome of 17 February was concluded, the men of *Discovery II* then immediately settled into a period of calm as routine work of every kind took precedence over all else. Hill made the customary journey to the solicitors' office, required after any return from a voyage, to deposit the usual Note of Protest about possible damage to *Discovery II.* This included not only any damage still not identified that they might have sustained in collision with pack ice, but also any damage to scientific gear caused by heavy weather. The Melbourne firm of solicitors, Stawell and Nankivell, gave Hill another resounding welcome.

He then went on to meet John King Davis once more. If he had had any trepidation about this, he need not have worried. This was a comfortable reunion. The two sailors met with broad smiles on their faces. Hill thanked Davis for his advice on the best way of working a safe path through the ice. They had refrained from pushing their way forward in the time-honoured manner of the old wooden ships. Never once did they intentionally butt the ice. Instead, they had simply nudged the obstruction in order to find the weakest point for entry, then gently eased the floe into two parts to create a lead between them. Davis' advice may well have saved Hill endless trouble in his steel ship. Hill subsequently passed on this advice, along with his own accumulated experience, to the Discovery Committee in London and recommended it as a procedure for those who would follow him.

There were other formal matters for Hill to arrange, so their first meeting did not last long. However, Davis made sure that the two of them met on more than one occasion while *Discovery II* remained in Williamstown Harbour. Davis introduced Hill to various friends and acquaintances, taking him to his club where he made him a Life Member. All the early recriminations were well forgotten, as Davis and Hill developed a pleasant friendship. In his April letter in April to Edward Nattriss, Hill remarked that, regardless of their early difficulties, he now had a warm, if qualified, regard for the older man.[3]

There were other practical matters for Hill to arrange. He sent Davis a copy of his Letter of Proceedings which covered the period from 1 January 1936 to 16 February, along with a chart showing the track of the vessel's journey during her voyage to and from the Bay of Whales. He also forwarded official photographs to the Department of Commerce and sent similar material to the Prime Minister of New Zealand. Even when still in the pack ice, Hill had received full instructions about the procedure for dealing with film and prints.[4] On 18 February he sent the Discovery Committee his Letter of Proceedings, reels of film and the official photograph album. The Discovery Committee had been anxious to know details of the list which would eventually reach them.[5] Hill warned the Committee that, due to the cold conditions, Saunders, the photographer, experienced considerable difficulty with the camera in taking the photographs. The technical quality, he assured them, was probably good. The photographs, superb by any standard, showed scenes from the start of the voyage with shots of the stores being loaded onto the ship at the Williamstown dockside, the poling party, the ice they encountered on the way and the aeroplane as it took off. Included were records of the standard scientific observations on plankton and hydrology in the neighbourhood of the Bay of Whales and the Ross Ice Barrier, and a note on the arrival of *Wyatt Earp*. There were several images of Little America taken from different angles, as well as crucial records of Ellsworth and Kenyon on board *Discovery II*.[6] When the film arrived in London, it was welcomed with considerable interest and relief as proof of the difficulties and successes of the operation. It also confirmed the Discovery Commission's copyright of the material.

Work had already begun to restore the ship to her normal condition. The ship's staff saw to it that they returned what remained of Commonwealth Government stores given to them before they set out. They also returned the aids lent to them by Sir Douglas Mawson and brought back aboard the scientific gear carefully stored ashore before the voyage south. Dockyard officials joined in the work of restoring *Discovery II* to her former state. A number of minor defects had become apparent during the recent voyage. These were small but required immediate attention before the ship could go to sea again. Numerous repairs due to wear and tear were completed. The Samson post, so essential for hauling nets and scientific gear overboard and inboard, was restored to its normal position. Decks, seriously marked by the oil drums, were scrubbed clean. Holes where nails had been driven into the deck to secure the aeroplane wings were filled and covering plates installed over them. The whaler was returned to its position over the hospital, the motorboat and its chocks to their normal place. The ship was repainted.

It had been startling to see how many alterations had been necessary before the rescue expedition left Melbourne. Now, in equal measure, it was satisfying to see the ship as she returned to her normal state. Hill submitted to the Discovery Committee a report of defects which Walker had drawn up, and listed the repairs as they were completed. He also made sure that dockyard officials, especially the overall manager, one Mr Eadie, were acknowledged for the help in speeding up the repair work.

He also sent the Discovery Committee a firm letter referring specifically to the motorboat, which he described as 'generally unreliable, commanding little confidence and being a danger to the ship's personnel'. The episode in the Balleny Islands had provided one of the most alarming moments of the entire voyage. On that occasion it had been crucial to recall scientists from a trip ashore because of heavy snow. Not for the first time, he wrote, had the

12 hp motorboat failed to start, and it fell to the pram dinghy around half its size to tow the 18-foot part-decked boat back to *Discovery II*. Hill's crisp tone in this letter clearly marked a new confidence in his role as commanding officer.[7] He felt keenly, as would any ship's master, his responsibility for the men under his command. For scientists and sailors to have been at such serious risk when the boat failed in the Balleny Islands struck him deeply. As a result of his protest, a new impulse starter was fitted when *Discovery II* reached Cape Town.[8]

By 29 February the dockyard employees had finished their work and the vessel was in all respects restored to her normal condition. The Lloyd's Surveyor inspected the ship and granted them the requisite Lloyd's Certificate of fitness to sail. *Discovery II* moved from Williamstown pier at 07.38 on 2 March and headed towards the oiling berth at Princes Pier, Port Melbourne, before taking on sufficient oil for the next cruise. After some delays while further repairs were done, first to the steering gear communication and then to the transformer on the shortwave receiver, they finally passed Port Phillip Head on 4 March. From there they set a course to clear the north-east point of Tasmania and head for 146° E. 44° 40′ S, then make for the edge of the pack ice off the Adelie coast of Wilkes Land.[9]

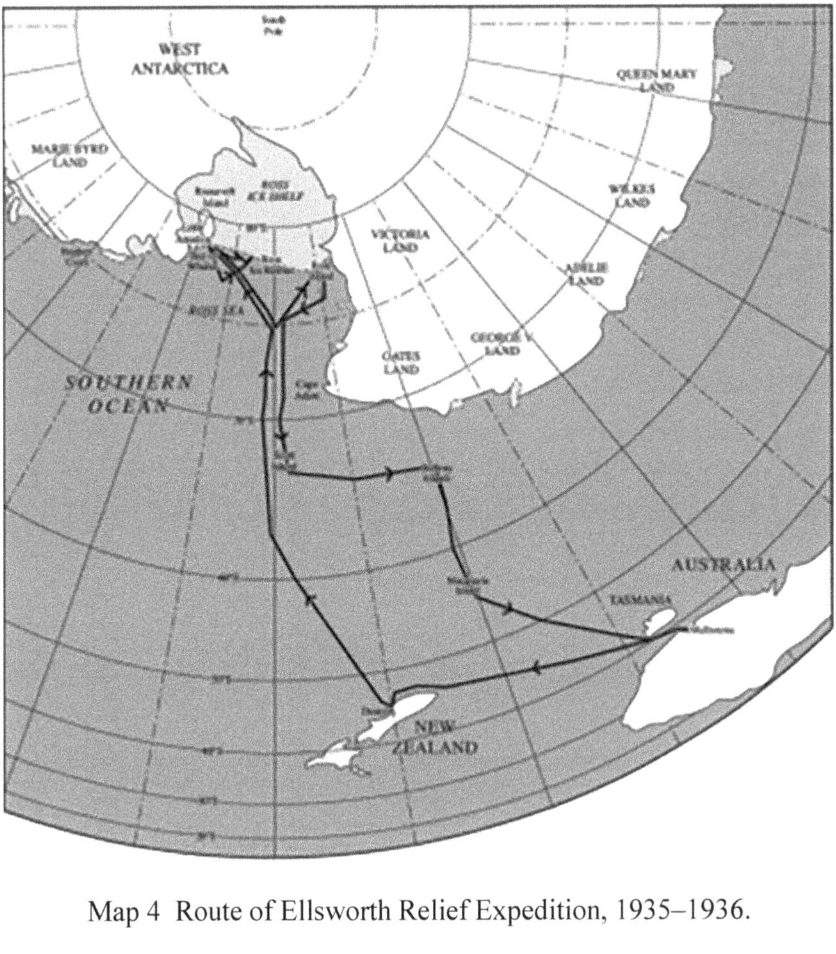

Map 4 Route of Ellsworth Relief Expedition, 1935–1936.

Chapter 32 EPILOGUE: A NORMAL CRUISE

During the speeches at their reception in Williamstown Hill played down the unusual nature of their recent voyage. In his words, the rescue of Lincoln Ellsworth and Hollick-Kenyon had just been a normal cruise, a term which in the 21st century suggests something leisurely. But there was never anything leisurely about any of the cruises *Discovery II* made, whether to or from the ice edge or along the ice edge itself, as scientists, helped by marine staff, worked long hours on the stations which underpinned the research of the Discovery Committee. Not just a cruise, as their expeditions were known, in the sense of something leisurely, these journeys meant extreme hard work for all concerned: for everyone involved in working through long hours on stations in temperatures below freezing point, for all who hauled the samples from the ocean and brought them aboard, and for the scientists who then worked in their small laboratories for even longer hours, below deck, in a small ship which pitched and tossed most of the time. For the marine staff it meant usually challenging navigation and the need for a special technique for handling the ship to enable the research.

The voyage to the Bay of Whales had been far from normal. It was certainly not normal for *Discovery II* to carry two aeroplanes, the Moth and the Wapiti, and to have its group of Australian airmen on board. The planes created problems for stowage, while the extra personnel on board meant seriously cramped accommodation for regular staff and crew, as well as incomers.

The cruise was also far from usual in its marine observations. Whereas *Discovery II*'s work was usually carefully planned to enable and enhance its scientific research, this expedition was totally unexpected and prevented the scientists from carrying out their intended programme. To this they never returned in the way so carefully prearranged.

However, though the circumstances were unusual, much good emerged. Most important was the main outcome of the expedition so suddenly re-planned. By his presence on the British research ship, Flight Lieutenant Eric Douglas was able to locate the two American aviators they had set out to find and so complete the joint British and Australian rescue mission. *Discovery II*'s voyage to The Bay of Whales enabled Ellsworth and Kenyon to find release from the long hours in their semi-underground confinement at Little America earlier than the two Americans had anticipated.

Ellsworth expressed full thanks to the Australian Government, the British Government, the Discovery Committee, and to Lieutenant Hill and his ship, *Discovery II*, as well as to the New Zealand Government for their part in providing a base from where *Discovery II* could set out. In his inscription to a copy of his book, *Beyond Horizons,* he presented to Hill on its publication in 1938, Ellsworth personally acknowledged Hill's efforts in carefully worded thanks for what he called 'an unforgettable voyage from the Bay of Whales to Australia in 1936'.

However, to be found by anyone other than Wilkins on *Wyatt Earp* was far from what Ellsworth had carefully planned. A few days more and their intended rescue ship, *Wyatt Earp*, would have appeared in the Bay of Whales. Ellsworth was disappointed not to meet up with Wilkins as he had originally intended, and to return to the waiting world unaided. What he may not have appreciated was how close his infected foot had led him to serious illness and to a far graver outcome had it not been treated when it was.

There were many other positive results from *Discovery II*'s unexpected cruise. In all his flights, whether to look out for ways through the pack ice or to find the Americans, Douglas showed remarkable flying skills. As a result of this mission, he was able to make valuable additions to recommendations for aerial reconnaissance in the difficult conditions of the Antarctic. Douglas was one of the earliest pioneers in this type of flying, while Hill and the Discovery Committee were completely new to it. Their joint report would prove invaluable for future work.[1]

Unusual though the push had been through the Ross Sea pack, *Discovery II*'s cruise showed new ways of taking a steel ship through thick ice. The ship had been threatened by danger from many quarters but on each journey through Ross Sea pack emerged un-holed and with her rudder undamaged. During this exercise Hill had the unstinting support of the entire ship's personnel, to which he added his own experience and knowledge of navigating in ice-infested waters. He was indebted to the Chief Scientist, Deacon, for his friendship, his wisdom, and his support. He was also able to draw on Captain John King Davis' advice and recommendations for navigation in ice conditions. It helped that the ice was never pressure ice in the way of Weddell Sea ice. However, any change in wind force or direction, or in the movement of the swell, could have caused irreparable damage to the ship and to the loss of its inmates. The risk was always a strong possibility. Again, Hill's reports and accounts proved invaluable for those who followed him.

An unusual cruise in so many ways, *Discovery II*'s voyage to The Bay of Whales was without doubt a normal cruise for them in one remarkable sense: the control of the expedition from over 10,000 miles away by a committee of men well distanced from the day-by-day trials of a small ship in often notoriously wild waters. Hill's position as Master of *Discovery II* was always delicate. It was part of his nature and upbringing to respect authority and to accept the advice of competent seniors. However, there were many occasions when, as did the scientists, he found the control of a Committee in Whitehall as demanding as the situation on the ground. Hill was always conscious of his responsibility to the others on board, 'to bring them back', as he would later say. On this occasion he also felt a firm sense of duty towards Douglas and the Australian airmen. Naturally, he felt strongly the need for clearer support from Whitehall.

Very few of the Committee members had real experience of the conditions in which their research ship worked. They had careers and responsibilities of their own, often very different from deep-sea or ice navigation. Importantly, the Discovery Committee was always conscious that they were accountable to the Colonial Office and to the Treasury for the public money they spent. Disregard for this would have imperilled the future of the research with which they were entrusted. In fairness to them, their own position and their accountability were not enviable. There were many instances of their tolerance and care for the men they sent out. Nevertheless, it seems extraordinary that they should have given priority to a discussion about the future of their iconic wooden ship *Discovery* over the mission of the moment, and to their need to remain silent about considerations of increased insurance premiums.

Some members had doubted Hill's courage when he had asked that the Ice Regulations be modified if the Committee required him to go deep into ice-infested waters. Harshly, it was even suggested that he seemed to be taking his instructions in 'a rather exaggerated way' and should be 'heartened up', as if he were being over-timid, when all Hill wanted was some recognition of the unusual nature of this voyage and acknowledgement that he was, seemingly, being required to contravene existing regulations. Though they had known it was still too early to send a ship into heavy ice and would have preferred a wooden ship to one of steel for the enterprise, they had refused both to grant the support Hill hoped for and to explain their reason, fear of infringing the terms of insurance policy. *Discovery II*'s success will have relieved them beyond measure.

Attitudes in London rapidly changed once news arrived that Ellsworth and Kenyon had been found alive and well. In London, members of the Discovery Committee were then prompt to recognise the extraordinary nature of *Discovery II*'s voyage through pack ice. The Committee were duly appreciative of Hill's reports and of the manner with which he had conducted the operations. They liked the way he acknowledged how well the entire Ship's Company had pulled together and the way in which the Australian airmen had been so supportive. They fully recognised Hill's leadership skills and management of what, they realised, had at times had been an alarming situation. Their thanks to the Commanding Officer were formal, but heartfelt.[2]

At first, the Committee thought the Polar Medal would be the most appropriate award for Hill, but after some research they realised that such medals were granted not just to one but to all members of a Polar expedition. On this occasion, they intended that Hill alone should be singled out for an award because of his particular contribution. In their submission that he should be appointed as an Officer of the Civil Division of the Order of the British Empire (OBE), the Committee asked that he should receive it at the earliest opportunity, 'forthwith if at all possible'.[3] Without doubt Hill had fulfilled the political aim of the Colonial Office and maintained the United Kingdom's standing as a presence in the Antarctic. It was, however, his overall leadership and navigational skills the Committee wished to emphasise. The award was gazetted on 14 February, and Hill subsequently received the insignia in the grounds of Government House, Melbourne, when they returned later that year.

The Polar Medal (Bronze) was awarded later, in 1942, to all who had served in *Discovery II* during her five commissions from 1929 to 1939.

The Committee recouped their losses in various ways. Even before *Discovery II* returned to Melbourne, a request came from Pathecine, representing Paramount News in London and New York, for the chance to secure good film of Hill, Ellsworth and the others involved, once they arrived in Australia. The film company also wanted the authority to oversee the negatives of the shots taken of rescue scenes.[4] It was a foretaste of the interest to come. Hill replied that he saw no objection to this but advised the film company to negotiate directly with the Discovery Committee in London.

On arrival in Melbourne, Hill ensured that the Governments of Australia and New Zealand received full sets of film and photographs taken.[5] The Committee then moved swiftly to establish control of the distribution of film for general publicity. All the countries involved were anxious to be able to have the footage as soon as possible. The Treasury had asked that the film rights be disposed of on a full-profit basis and the Committee quickly claimed their rights to film and other photographic material. They had already reached an understanding with the Governments of Australia and New Zealand Governments that they and the other two Governments were to be supplied with the film, and each Government was to be at liberty to market it as they thought fit. The Committee would now control all other outlets.

In the United Kingdom news of the rescue continued to cause great excitement for several months. It was encouraging for people to read not only of the remarkable flight made by the two Americans, but also to realise that the British nation was capable of such an exploit as leading the rescue party. It was a strong antidote to feelings that the British Empire was dwindling in size and extent. It also provided a distraction from concerns about the rise of Nazism in Germany.

The Committee was also anxious to correct the great amount of erroneous matter that had been published in the press and thought it important for a reliable account now to be presented. Most importantly, they felt this would be of valuable assistance to ships in the future when navigating in similar circumstances.

The committee spread publicity of the rescue as far as they could. They took Hill's Forty Ninth Letter of Proceedings, the relevant report submitted by the captain, as the ideal piece to offer *The Times* for publication as a special article. It gave an excellent account, the Committee felt, of the handling of the ship in the ice. Used as a piece for *The Times*, it might also be of general interest to the public as a whole. If it were not to be given to *The Times*, to whom the Colonial Office could not give sole rights, it should be offered to *The Marine Observer*. In this way it would be available to all navigating officers for its account of the seamanship involved. However, *The Marine Observer*, the quarterly journal of the Meteorological Office, preferred a direct contribution from the Commanding Officer himself.[6] A briefer version would be offered to *The Polar Record*, the regular Cambridge journal of Arctic and Antarctic research published by the Scott Polar Research Institute.

The Committee decided that the piece should carry the by-line of Sir Percy Douglas rather than Hill's. In their opinion an account written by an admiral would carry more weight than that of the younger officer who had been the captain of the rescue ship in question, however junior in rank. The article appeared duly in the July issue, 1936, as written by Sir Percy, but fully based on Hill's Forty Ninth Report of Proceedings.[7] The *Polar Record* fully acknowledged the source of the account, but it remained a curious bias in the Committee's thinking that a senior officer's by-line would bear more authority.

Meanwhile, those on board *Discovery II* had little idea of the details being discussed by the Discovery Commission in London. Admittedly, someone, whether wit or cynic remains unknown, had written the words 'Saviours of Empire' on the reverse of a small photograph of the ship's officers. Clearly those on board *Discovery II* realised that diplomatic relations and politics had played a large part in the response of the Colonial Office and the Discovery Committee to the Australians' request for help. But the finer details under negotiations were, in general, not their concern. While matters of publicity and recompense remained of the utmost concern to some, their research ship in the Southern Ocean soon returned to its regular routine. Their work, to an extent little known at the time, made a crucial contribution to an understanding of the Southern Ocean, to hydrography and to the ultimate protection of whales.[8]

When they first set off from Williamstown, many on board missed the challenge and excitement their rescue mission to 74° S had given them. The absence of the cheerful camaraderie of the Australian airmen left another gap in their lives on board. But they were pleased to get back to their real work, the day-to-day navigation and research of their commission. Their next course was to brave the westerly gales yet again, this time far worse than even those they had experienced on the way north from the Ross Sea to Melbourne. After several delays when they were forced to lie hove to, they gradually made for the ice edge. They met the pack ice on

16 March and for ten days worked their way steadily westward along the Antarctic coast, before turning north through yet more of the almost inevitable westerlies to reach Fremantle on 7 April.[9] Well aware by now that they had been meeting political demands, they were now back to what they knew as a normal cruise: long days of hard work, sometimes in hot sunshine, sometimes with heavy snow and minimal visibility, but always in an environment which provided them with memories of places of exquisite beauty laced with many moments of extreme danger.

Sources and References

Abbreviations

NLS. National Library of Scotland, Edinburgh.
NOL. National Oceanographic Library, Southampton.
SPRI. Scott Polar Research Institute, Cambridge.

Chapter 1: Discovery II at the Ice Edge

1 NOL MS 4/3/1/60, R.R.S. *Discovery II Ship's Log*, 3rd October 1935
2 NLS MS 9558/232-236, *Programme of R.R.S. Discovery II, 1935-1936.*
3 Deacon, G.E.R., J.W.S. Marr and G.W. Rayner, The Antarctic Voyages of R.R.S. Discovery II and R.R.S. William Scoresby, 1935-37, *The Geographical Journal,* vol. 93, No. 3, 1939, 188ff. https://doi.org/10.2307/1788354
4 Haddelsey, Stephen, *Ice Captain: The Life and Times of J.R. Stenhouse,* The History Press, 2008, 132-133.
5 Marsden, R. 'Expedition to Investigation,' *Understanding the Oceans,* ed. Margaret Deacon, Tony Rice and Colin Summerhayes, UCL Press, 2001, 81-82.
6 Mackintosh, N.A. The Work of the Discovery Committee, *Proceedings of the Royal Society of London. Series A, Mathematical and Physical Sciences,* 202, no.1068, 1950, 137.
7 Marsden, 'Expedition to Investigation', 81-82.
8 Hardy, A. C., Whale-Marking in the Southern Ocean, *The Geographical Journal,* vol. 96, no. 5, 345-50. https://doi.org/10.2307/1788808
9 Ardley, R.A.B., RNR and Mackintosh, N.A., DSc., The Royal Research Ship *Discovery II, Discovery Reports, vol. xiii,* Cambridge University Press, July 1936, 81.
10 Ardley and Mackintosh, 82ff.
11 Mackintosh, N.A., The Third Commission of the R.RS. *Discovery II, The Geographical Journal,* vol. 88, No. 4, 1936, 306.111 304-318. https://doi.org/10.2307/1786334
12 *The Times,* December 31, 1930.
13 NLS MS 9554/195 (Ref.137054/31/MO1), Letter from Marine Superintendent, Meteorological Office to Secretary Discovery Committee, August 9, 1933.
14 Marsden, 'Expedition to Investigation', 81-82.

Chapter 2: Fourth Commission

1 NLS MS 9558/279-280; SPRI MS 1284/8; CC, 145th meeting, Discovery Committee, September 16th, 1935; Ardley, and Mackintosh, 81.

2 NLS MS 9560/249; SPRI MS 1284/5; CC, 151st Meeting, Discovery Committee, May 1st, 1936.

3 *Programme of R.R.S.* Discovery II, *1935-6.*

4 Deacon, G.R., A General Account of the Hydrology of the South Atlantic Ocean, *Discovery Reports, vol. vii,* Cambridge University Press, November 1933, 173-238.

5 NOL MS 2/4/102, Letter to Lieutenant L.C. Hill, from Discovery Committee, August 1935; SPRI MS 1284/8;ER, Discovery Committee, Ship Subcommittee, August 1935; 145th Meeting, Discovery Committee.

6 SPRI MS 1284/9;ER, Discovery Committee, Personnel Subcommittee, October 7th, 1931.

7 NOL MS 2/4/37, Captain W. Carey, *Report on Services of Mr. L.C. Hill, in the* R.R.S. *Discovery II,* August 1932.

8 Thomas, L.H., *Unpublished Journal,* by courtesy of Leona J. Thomas.

9 NOL MS 2/5/89; SPRI MS 1284/8;CC, Discovery Committee, Ship Subcommittee, September 13th, 1935.

10 NLS MS 9558/238 ff, Borley, J.O., handwritten letter to J.M. Wordie, 23.9.35.

11 NLS MS 9558/239, Wordie, J.M. reply to J.O. Borley, September 24th, 1935.

12 NLS MS 9558/245, Borley, J.O., to J.M. Wordie, September 1935.

13 NLS MS 9558/245 Hill, L.C., to J.M. Wordie from R.R.S. *Discovery II,* November 1935.

14 NLS MS 9558/32, Vice-Admiral Sir E.R.G. Evans, Commander-in-Chief Africa Station, personal letter to Sir Percy Douglas, 28th January 1935.

15 SPRI MS 1268/5, Mackintosh, N.A., *Diary kept on board Discovery II, August, 1934-July, 1935.*

16 *Captain's Forty seventh Letter.*

17 NOL MS 2/5/90 Telegram no. 20 from Secretary, Discovery Committee to Master, *Discovery II,* 4th December 1935

18 *Captain's Forty seventh Letter.*

19 Douglas, Vice-Admiral Sir Percy, *A Meeting in Antarctica,* The Times, July 16, 1934.

20 NOL MS 2/3/459-8a (22690/26014), Telegram no 346 from Master, R.R.S. *Discovery II* to the Secretary of the Discovery Committee, 4th December 1935.

Chapter 3:Full Speed to Melbourne

1 NOL MS D21213, Polar Committee, Rymill Expedition Subcommittee 1934.
2 NOL MS 4/3/1/60, R.R.S. *Discovery II Ship's Log,* 5th December 1935; Thomas, *Unpublished Journal.*
3 Thomas, *Unpublished Journal.*
4 Thomas, *Unpublished Journal.*
5 *Captain's Forty seventh Letter;* NOL MS 2/3/459 Telegram to Director Navigation, Melbourne, December 1935.
6 *Captain's Forty seventh Letter;* NOL MS 2/3/460, Melbourne Herald to Captain, R.R.S. *Discovery II,* December 6th, 1935
7 *Captain's Forty seventh Letter;* NOL MS 2/3/459 (22690/29), Telegram from Master R.R.S. *Discovery II* to Melbourne Herald, December 11, 1935.
8 *Captain's Forty seventh Letter.*
9 SPRI MS 1510/9/3, Hill, L.C., Letter to E.A. Nattriss, 5th April 1936.
10 NOL MS 2/3/459 (22690/29), Telegram from Master, R.R.S. *Discovery II* to Director Navigation, Melbourne, December 6th, 1935.
11 Hill to Nattriss, 5th April 1936.
12 NOL MS 2/3/460 (22690/2), Telegram No 21 from Secretary, Discovery Committee, to Master, R.R.S. *Discovery II,* December 1935.
13 NOL MS 1/C160, Millard, H., to J.M. Troutbeck, December 7th, 1935.

Chapter 4: Lincoln Ellsworth

1 Wilkins, Hubert. The Wilkins-Hearst Antarctic Expedition, 1928-1929, *Geographical Review* 19, no. 3 (1929), 353-376. https://doi.org/10.2307/209145
2 Mill, Hugh Robert, The Significance of Sir Hubert Wilkins' Antarctic Flights, *Geographical Review* 19, no. 3, 377-386; Wordie, J. M., Sir Hubert Wilkins' Discoveries in Graham Land, *The Geographical Journal* 73, no. 3, 254-57. https://doi.org/10.2307/209146
3 Wilkins, Hubert, Further Antarctic Explorations, *Geographical Review* 20, no. 3, (1930), 357-88. https://doi.org/10.2307/209099
4 Byrd, *Little America, Aerial Exploration in the Antarctic: The Flight to the South Pole,* G.P. Putnam's Sons, New York, London, 1930, 92.
5 Saunders, Harold E., The Flight of Admiral Byrd to the South Pole and the Exploration of Marie Byrd Land, *Proceedings of the American Philosophical Society,* vol. 82, no. 5, 1940, 801-820. http://www.jstor.org/stable/984892
6 Mawson, Douglas, The British, Australian and New Zealand Antarctic Research Expedition (BANZARE), 1929-31, *The Geographical Journal,* vol. 80, No. 2, (1932), 101-126.

7 Ellsworth, Lincoln, *Beyond Horizons,* William Heinemann, Ltd, London & Toronto, 1938, 235-6.

8 *The Times,* November 21, 1935.

9 *The Times,* November 25, 1935.

Chapter 5: Silence

1 Ellsworth, Lincoln, My Flight Across Antarctica, *The National Geographic Magazine, The National Geographic Society,* (July 1936), 8.

2 Ellsworth, Flight, 1-35; Ellsworth, Lincoln, The First Crossing of Antarctica, *The Geographical Journal* vol. 89, no. 3 1937, 193-209. https://doi.org/10.2307/1785792

3 Ellsworth, Flight, 26-31.

Chapter 6: Rescue Alert

1 e.g. 'Ellsworth Not Reported', *News (Adelaide), SA,* Wed, Nov 27 1935,1, Trove online database. https://trove.nla.gov.au/help/categories/newspapers-and-gazettes-category.

2 The Times, November 25, 1935.

3 Trove online database. *The Western Australian of Perth,* National Library of Australia, November 28, 1935.

4 'Radios From Ellsworth (Wellington, N.Z.)', *The Daily Telegraph (Sydney), NSW,* Fri, Nov 29, 1935, Trove online database.

5 *The Times,* November 29, 1935.

6 *The Times,* November 25, 1935.

7 NOL MS 2/3/460 (22690) from the Prime Minister of the American Commonwealth of Australia to the Secretary of State for Dominion Affairs. December 2, 1935, No 90.

8 *The Times,* November 25, 1935.

Chapter 7: Reactions in London

1 NOL MS 2/5/90, Discovery Commission Emergency Meeting, December 3rd, 1935.

2 Douglas, Vice-Admiral Sir Percy, A Meeting in Antarctica, *The Times,* July 16, 1934.

3 Lintott, Bryan, *The British Graham Land Expedition 1934-37,* Scott Polar Research Institute, 2010.

4 NLS MS 9559/8-11, Letter from Borley, J.O. to J.M. Wordie, December 5th, 1935.

5 Letter from Borley, J.O. to J.M. Wordie, December 3, 1935.

6 NOL MS 1/C160 Troutbeck, John, *Memorandum from Foreign Office to H. Millard, American Embassy,* December 4th, 1935.
7 *The Times,* December 4, 1935.

Chapter 8: Action Stations

1 *The Times,* December 3, 1935.
2 NOL MS 4/3/1/65, Millard, Hugh, American Embassy to J.M. Troutbeck, Foreign Office, December 3rd, 1935.
3 NOL MS 1/C160, Troutbeck to Millard, December 4th, 1935.
4 NOL MS 2/5/90, Correspondence between Hugh Millard, American Embassy, and J.M. Troutbeck, Foreign Office, London SW1, December 1935.
5 NOL MS 2/5/90, Discovery Commission Emergency Meeting, December 9th, 1935.
6 Millard to Troutbeck, December 5th, 1935; Troutbeck to Millard, December 4th, 1935; Millard to Troutbeck, December 7th, 1935.
7 *The Times,* December 6, 1935; Message circulated at Emergency Meeting, December 9th, 1935.
8 *The Times,* December 9, 1935.
9 Millard to Troutbeck, December 5th, 1935.
10 *The Times,* December 9, 1935; NOL MS 1/C160, Millard to Troutbeck, December 10th, 1935.
11 NOL MS 1/C160, Rymill, J., c/o American Embassy, to Hugh Millard, December 14th, 1935.

Chapter 9: Many Masters

1 NOL MS 2/5/90, Correspondence between Hugh Millard, American Embassy, and J.M. Troutbeck, Foreign Office, London SW1, December 1935; NOL MS 1/ C160, Wilkins, Sir H. from *Wyatt Earp* to the US Embassy, December 7th, 1935, forwarded to Discovery Committee and received, December 9th, 1935.
2 NOL MS 23460 (22690), Telegram no 21, From Secretary, Discovery Committee to Master, *Discovery II,* December 11th, 1935.
3 *The Times,* December 9, 1935.

Chapter 10: Discovery II: Her Ice Experience

1 Byrd, *Little America,* 77-97.
2 NOL MS 2/3/170, 'Whaling in the Antarctic', *Manchester Guardian,* August 4, 1933.

3 Mackintosh, N.A., DSc, and H.F.P. Herdman, Distribution of the Pack Ice in the Southern Ocean, *Discovery Report, vol. xix,* 1940, 285-296.

4 Ross, James Clark, *A Voyage of Discovery and Research in the Southern and Antarctic Regions, during the Years 1839-43, vol. 1, London: John Murray, 1847.*

5 Priestley, Raymond E., The Work and Adventures of the Northern Party of Captain Scott's Antarctic Expedition, 1910-1913, *The Geographical Journal* 43, no. 1, 1914, 1–14.

6 Haddelsey, *Ice Captain,* 53 ff.

7 Ardley, and Mackintosh, 81-85.

8 Thomas, *Unpublished Journal.*

9 NOL MS 2/3/326 (11274) *Captain's 48th Letter of Proceedings for R.R.S. Discovery II from 1st October to 4th November 1930*, Dunedin, 31st December 1935.

10 *NOL MS 4/3/1/60,* R.R.S. *Discovery II Ship's Log,* 17th November 1930.

11 Thomas, *Unpublished Journal.*

12 NOL MS 2/3/3326 (11274) Carey, W., *Report of Proceedings for R.R.S. Discovery II,* 4th November 1930.

13 Kemp, Stanley. The Voyage of the R.R.S. *Discovery II*: Surveys and Soundings, *The Geographical Journal* 79, no. 3, 1932, 168-81. https://doi.org/10.2307/1785191

14 NOL MS 2/3/339 (12192), Carey, W., *Captain's Report of Proceedings for R.R.S. Discovery II,* December 1931.

Chapter 11: Forever Ice

1 John, D. Dilwyn, The Second Antarctic Commission of the R.R.S. *Discovery II, The Geographical Journal* 83, no. 5, 1934, 381-394. //doi.org/10.2307/1785722

2 Thomas, *Unpublished Journal.*

3 John, Second Antarctic Commission, 386

4 SPRI, MS 1284/4/17/1752;CC, Minutes and Related Papers.

5 NOL MS 2/3/339 (12192), Carey, W. *Captain's Eleventh Report of Proceedings, February 1932*;.

6 Mackintosh, *Diary.*

7 NLS 9558/23-236, *Programme of* R.R.S. *Discovery II 1935-1936*, 2.

Chapter 12: Ice Regulations

1 NLS MS 9549/164-165ff. Messrs Glanvill, Enthoven and Company to Sir Fortescue Flannery, February 10th,1931.

2 NLS MS 9549/166ff. Sir Fortescue Flannery to Messrs Glanvill, Enthoven and Company, April 29th, 1931.
3 NLS MS 9549/161-162, Messrs Glanvill, Enthoven and Company to the Secretary, Discovery Committee, June 1st, 1931.
4 SPRI MS 1284/8;CC, Minutes, Discovery Committee, Ship Subcommittee meeting, June 30th, 1931.
5 SPRI MS 1284/4/15;CC, Minutes, Discovery Committee Main Committee, 106th Meeting, July 2nd, 1931.
6 NLS MS 9549, 181, 182, Darnley E.R. (Chairman to Discovery Committee) to J.M. Wordie, July 7th, 1931.
7 Cf. NLS MS 947/311-320, Wordie correspondence with S.F. Harmer (Vice-Chairman, Discovery Committee) and elsewhere.

Chapter 13: Melbourne

1 NOL MS 4/3/1/60, R.R.S. *Discovery II Ship's Log,* 12th December 1935; NOL MS 2/3/492, Telegram 356 from Master R.R.S. *Discovery II* to Secretary, Discovery Committee, December 13th, 1935; *Captain's 48th Letter.*
2 SPRI MS 1510/9/1, Bainbridge, S.A., to Captain H. Willey, re report of *Discovery II* Ellsworth expedition, February 21st, 1936.
3 *Captain' 48th Letter.*
4 Bainbridge, S.A., to Captain H. Willey.
5 Thomas, *Unpublished Journal.*
6 Telegram 356, to Discovery Committee.
7 NOL MS 2/3/492, Telegram no 24 from London to R.R.S. *Discovery II,* December 11th, 1935.
8 NOL MS 2/3/492 (22690/2), Telegram no 25, from Secretary Discovery Committee to Captain, R.R.S. *Discovery II,* December 19th, 1935.
9 *Captain's 48th Letter.*
10 NOL MS 2/3/492 (22690), Telegram no 41 from Captain, R.R.S. *Discovery II* in reply to Telegram 24, December 17th, 1935.
11 *Daily Telegraph,* December 21, 1935.
12 NOL MS 2/3/492 (22690) 35, Telegram no 34 from Secretary, Discovery Committee to Master, R.R.S. *Discovery II,* December 20th, 1935.

Chapter 14: To Business

1 Davis, J.K., *With the 'Aurora' in the Antarctic, 1911-1914,* London: Andrew Melrose, Ltd, Nabu Public Domain Reprints, 53-54.
2 *Manchester Guardian,* February 27, 1914.

3 Hill to Nattriss, 5th April 1936.
4 NOL MS 2/3492 (22690/2), Telegram no 357 from Captain, R.R.S. *Discovery II* to Secretary, Discovery Committee, December 16th, 1935.
5 Ardley, and Mackintosh, 81.
6 NOL MS 2/3/492 (22690/2), Telegram no 24, December 17th, 1935.
7 NOL MS 2/3/492 (26335), Telegram 34 from Master, R.R.S. *Discovery II* to Secretary, Discovery Committee, December 16th, 1935.
8 NOL MS 2/3/492 (26335), Telegram no 24 from Discovery Committee to Master, December 17th, 1935.
9 Hill to Nattriss, 5th April, 1936.
10 *Captain's 48th Letter.*
11 *Captain's 48th Letter.*

Chapter 15: A Full Stow

1 NOL MS 22690/2, Hill, L.C., *Report made to the Discovery Committee in co-operation with Flight Lieut. Douglas, R.A.A.F.,* February 1936.
2 Hill and Douglas, *Report.*
3 Hill and Douglas, *Report.*
4 SPRI MS 1510/8/50, Hill, L.C. to E.A. Nattriss, 31st December, 1935; Ommanney, F.D., *South Latitude,* The Longman Library, 1940, 186-8.
5 NOL MS 1/C160, From Commonwealth of Australia, Department of Defence, Melbourne, to Capt. L.C. Hill, Officer Commanding, Ellsworth Relief Expedition, R.R.S. Discovery II, 21.12.35.
6 *Captain's 48th Letter*
7 NLS MS 9559/176-178, Davis, J.K., *Notes re Passage of Ice Pack into Ross Sea,* Melbourne, December 1935.
8 NLS MS 9559/180-182, *Preliminary Outline of Procedure to be Adopted by Discovery II,* Melbourne, December 1935.

Chapter 16: Through the Bass Strait to Dunedin

1 NOL MS 2/3/492 (26335), Telegram 362, From Master, R.R.S. *Discovery II* to Discovery Committee, December 24th, 1935, received 10.45, December 25th, 1935.
2 *Captain's 48th Letter*; Deacon, 1939, 189.
3 NOL MS 4/6/1/62, R.R.S. *Discovery II Ship's log,* 25th December, 1935.
4 Bainbridge, S.A., to Captain H. Willey..
5 *Captain's 48th Letter.*
6 *The Times,* December 31, 1935.
7 SPRI MS 938/2, Douglas, Flight Lieut. Eric, *Rough Log of Voyage to the Bay of Whales Ross Sea made in* R.R.S. *Discovery II to locate Lincoln*

Ellsworth and Hollick-Kenyon after their trans-Antarctic flight Jan. 1936, Douglas, 2.

8 *Captain's 48th Letter.*

Chapter 17: Pending

1 NOL MS 2/3/492 (26335) Telegram 356, from Hill to Borley, December 16th, 1935.

2 NLS MS 9559/158 ff., letter from Borley, J.O., to J.M. Wordie, January 5th, 1935.

3 NOL MS 2/3/492 (26335), Telegram no 386 from Discovery Committee to Master, R.R.S. Discovery II, December 16th, 1935.

4 NOL MS 2/3/459-71, letter from Wiseman, R.A. to F.T. Sandford, January 1st, 1936.

5 SPRI MS 1248/4;CC, Discovery Committee, Ship Subcommittee, December 11th, 1935.

6 NLS MS 9559/85-88, 'Future of R.R.S. *Discovery*', letter from J.M. Wordie to Mr Harper, October 19th, 1934.

7 SPRI MS 1284/5, Special Subcommittee meeting, December 17th, 1935.

8 NOL MS 9559/143; letter from Borley, J.O. to J.M. Wordie, December 27th, 1935.

9 NOL MS 2/5/90, Telegram from Master, *Discovery II* to Discovery Committee, December 4th, 1935; NOL MS 2/3/459-64,0, from Discovery Committee to Master, R.R.S. *Discovery II*, December 30th, 1935.

10 NLS MS 9559/162, letter from Deacon, G.E.R. to Dr S. Kemp, December 23rd, 1935.

11 Cf SPRI MS 1284/8/2, Ship Subcommittee, July 25th, 1930.

12 NLS MS 9559, letter from Borley, J.O. to J.M. Wordie, January 8th, 1936.

13 NOL MS 2/3/492, Telegram no. 3 to Master, R.R.S. *Discovery II*, January 9th, 1936.

Chapter 18: Brief Respite

1 PC, to BMC.

2 Hill, to Nattriss, 5th April 1936.

3 *Captain's 48th Letter*; Deacon, 1939, 189.

4 SPRI MS 15/10/9/3, Hill, L.C., to E.A. Nattriss, written from R.R.S. *Discovery II*, 31st December 1935.

5 NOL MS 2/3/496 (26465), *Captain's Forty Ninth Letter of Proceedings for R.R.S. Discovery II,* Melbourne, 16th February 1936.

6 Douglas, *Rough Log,* 2.

Chapter 19: Southern Ocean

1 *Captain's Forty Ninth Letter*; NOL MS 2/3/492, Telegram 372 to Discovery Committee, January 2nd, 1935.
2 Telegram 372 to Discovery Committee, January 2nd, 1935.

Chapter 20: Towards the Pack Ice

1 Douglas, *Rough Log,* 6.
2 Douglas, Vice-Admiral Sir Percy, Mr Ellsworth and the Discovery II, *The Polar Record,* no 12: July 1936, 156-172.
3 NOL MS 2/3/492, Telegram 378, Hill to Discovery Committee, January 5th, 1935.
4 NOL MS 2/3/492 (22690/29) Telegram 12 to Director Navigation, Melbourne, January 7th, 1935.
5 NOL MS 2/3/492, Telegram 153 from Discovery Committee, London to Captain, R.R.S. *Discovery II,* December 19th, 1936.
6 Davis, *Notes re Passage of Ice Pack.*
7 Davis, *Notes re Passage of Ice Pack.*
8 *Captain's Forty Ninth Letter.*
9 Douglas, Vice-Admiral Sir Percy, *The Polar Record*; Ommanney, *South Latitude,* 186.
10 NOL MS 102/3/502 (26870), Marr, J., *Biological Work: Dunedin to the Bay of Whales,* in Discovery Committee, Twentieth Scientific Report on the work of the Royal Research Ship *Discovery II,* December 24th, 1935-February 5th, 1936, Deacon, G.E.R., J.W.S. Marr, F.D. Ommanney.

Chapter 21: Following the Leads

1 Douglas, *Rough Log,* 3-4.
2 Ommanney, *South Latitude,* 186.
3 Davis, J.K., *Notes re Passage of Ice Pack.*
4 Davis, *Notes re Passage of Ice Pack.*
5 Haddelsey, *Ice Captain,* 132-3.
6 Mackintosh and Herdman, Distribution of the Pack Ice in the Southern Ocean, *Discovery Report, vol xix,* 285-296, 1940
7 Douglas, *Rough Log,* 6.
8 *Captain's Forty Ninth Letter.*
9 Douglas, *Rough Log,* 6.
10 Deacon, 1939, 189.

Chapter 22: Ice Bound

1 Saunders, A., *A Camera in Antarctica,* Winchester Publications Limited, 1950, 128.
2 NOL MS 2/3/492 (22690) Telegram No. 387, from Commanding Officer to Discovery Committee, London, January 11, 1935; NOL MS 9559/294, 148th Meeting of the Discovery Committee, Colonial Office, Tuesday, January 21st, 1936.
3 NOL MS 2/3/496 (26465), *Captain's Forty Ninth Letter of Proceedings for R.R.S. Discovery II,* Melbourne, 16th February 1936.
4 Deacon, 1939, 190.
5 Deacon, 1939, 190.
6 *Captain's Forty Ninth Letter.*
7 NOL MS 22690/2, Hill, L.C., *Report made to the Discovery Committee in co-operation with Flight Lieut. Douglas, R.A.A.F.,* February 1936.
8 Ommanney, *South Latitude,* 187.
9 Saunders, *A Camera in Antarctica,* 129.
10 Mackintosh, N.A., The Third Commission, 307.
11 Ommanney, *South Latitude,* 190.
12 Ommanney, *South Latitude,* 186-187.
13 Hill, and Douglas, *Report,*.
14 Hill and Douglas, *Report.*

Chapter 23: Help from the Air

1 *Captain's Forty Ninth Letter.*
2 *Captain's Forty Ninth Letter;* Ommanney, *South Latitude,* 189.
3 *Captain's Forty Ninth Letter.*
4 Hill and Douglas, *Report.*
5 Hill and Douglas, *Report.*
6 Douglas, *Rough Log,* 9.
7 NOL MS 2/3/492, 148th Meeting of the Discovery Committee, January 21, 1936; Telegram 388, Commanding Officer to Discovery Committee, January 12th, 1936; Douglas, *Rough Log,* 9.
8 Hill and Douglas, *Report.*
9 Douglas, Flight Lieut. Eric, *'Handwritten Report on Relief Flight to Mr Lincoln Ellsworth and Mr Hollick Kenyon for* R.R.S. *Discovery II,* January 1936; Douglas, *Rough Log,* 10.
10 Douglas, *Rough Log,* 10.
11 *Captain's Forty Ninth Letter.*
12 NOL MS 2/3/492 (22690/29), Telegram 388, from Captain, *Discovery II* to Discovery Committee, January 13th, 1936.

13 Douglas, Vice-Admiral Sir Percy, *Polar Record.*
14 Ommanney, *South Latitude,* 189-190.

Chapter 24: Into the Bay of Whales

1 NOL MS 2/3/492 Telegram no. 390 from Commanding Officer to Discovery Committee, January 14th, 1935; *Captain's Forty Ninth Letter*.
2 NOL MS 2/3/492 (22690) Telegram No 6 from Colonial Office to Master, R.R.S. *Discovery II,* January 13th, 1936.
3 NOL MS 9559/294, 148th Meeting of the Discovery Committee, Colonial Office, 21st January 1936.
4 NOL MS 2/3/492 (22690/2) Telegram 390 from Master, R.R.S. *Discovery II* to Secretary Discovery Committee, January 14, 1936.
5 NLS MS 9560/215-216, Wireless Report, from A.E. Morris, Wireless Operator, R.R.S. *Discovery II* to Commanding Officer, *Discovery II,* March 31, 1936, 6.
6 NLS, MS 5960/168 ff. SPRI MS 1284/8, Discovery Committee, Ship Subcommittee, 24th April 24, 1936.
7 NOL MS 2/3/501, *Captain's Fiftieth Letter of Proceedings,* Fremantle, April 7, 1936.
8 Douglas, *Rough Log,* 11.
9 Bainbridge, S.A., to Captain H. Willey.
10 Bainbridge, S.A., to Captain H. Willey.
11 Saunders, *A Camera in Antarctica,* 128.
12 Deacon, 1939, 189.
13 Marr, Twentieth Scientific Report, 1936.

Chapter 25: Found!

1 Douglas, *Rough Log* 13-15.
2 Douglas, *Rough Log,* 13-15.
3 The Times, February 18, 1936; Ommanney, *South Latitude,* 202.
4 Bainbridge, S.A., to Captain H. Willey.
5 Ommanney, *South Latitude,* 209-210.
6 Ellsworth, Flight, 26-31.

Chapter 26: 'Not "Rescued"-"Aided"

1 Ellsworth, *Beyond Horizons,* 359.
2 *New York Times,* June 1929.
3 Ellsworth, *Beyond Horizons,* 325.
4 Ellsworth, *Beyond Horizons,* 308.

5 Ellsworth, *Beyond Horizons,* 325.
6 Ellsworth, *Beyond Horizons,* 330.
7 Ellsworth, *Beyond Horizons,* 330.
8 NOL MS 26284, Medical Report for R.R.S. *Discovery II,* February 1936.
9 Ellsworth, *Beyond Horizons,* 330.
10 Douglas, Vice-Admiral Sir Percy, Completed Expedition: Ellsworth Antarctic Expedition, Mr Ellsworth and the *Discovery II, Polar Record,* no 12, July 1936.
11 *Captain's Forty Ninth Letter.*

Chapter 27: After the Rescue

1 NOL MS 2/6/430 (22690/2), Telegram to Master, *Discovery II,* 10th January 1936.
2 *The Times,* January 17, 1936.
3 *The Times,* January 18, 1936.
4 NOL MS 2/3/6/460 (22690), Replies to *Daily Mail* and *News Chronicle,* London.
5 NOL MS 2/5/90 Telegram 346, 4th December, 1935, Hill to Secretary Borley, with addendum by Deacon; NOL MS 2/3/459 -64, Telegram from Discovery Committee to Master, R.R.S. *Discovery II,* 30th December, 1935.
6 NOL MS 2/3/459-58, Telegram 366 from Chief Scientific Officer, *Discovery II* to Dr Kemp, Discovery Committee, December 28th, 1935.
7 NOL MS 2/3/459-64, Telegram no 34 to Master, R.R.S. *Discovery II,* December 30th, 1935.
8 NOL MS 22690 35, Telegram 43 to Chief Scientist, R.R.S. *Discovery II,* January 19th, 1936.
9 Deacon, Twentieth Scientific Report, 14, 29.
10 Deacon, Twentieth Scientific Report, 14.
11 Deacon, 1939, 190.
12 Deacon, Twentieth Scientific Report.
13 Douglas, *Rough Log,* 20.

Chapter 28: Wyatt Earp

1 Ellsworth, *Beyond Horizons,* 238.
2 Ellsworth, *Beyond Horizons,* 339.
3 Ellsworth, *Beyond Horizons,* 238.
4 Douglas, *Rough Log,* 22.
5 Bainbridge, S.A., to Captain H. Willey.
6 Douglas, *Rough Log,* 23.
7 Ommanney, *South Latitude,* 216.

8 Douglas, *Rough Log,* 23.
9 Douglas, *Rough Log,* 19.
10 Douglas, *Rough Log,* 22.
11 SPRI MS 1284/6/4 Scientific Subcommittee, December 16th, 1935.
12 Douglas, *Rough Log,* 24.

Chapter 29: Return through the Pack

1 Deacon, Twentieth Scientific Report, 121, 123.
2 Deacon, 1939, 191.
3 Deacon, Twentieth Scientific Report.
4 Deacon, 1939, 193.
5 Douglas, *Rough Log,* 41.
6 NOL MS 22690/226467, Douglas, Flight Lieut. Eric, *Report on Relief Flight to Mr. Lincoln Ellsworth and Mr. Hollick Kenyon Per R.R.S. Discovery II, January 1936.*
7 Hill and Douglas, *Report.*
8 SPRI MS 1284/5, 151st Meeting of Discovery Committee, May 1, 1936.

Chapter 30: Winding Down

1 Marr, J., Twentieth Scientific Report 193.
2 Ellsworth, *Beyond Horizons,* 339.
3 Douglas, *Rough Log,* 242.
4 Ommanney, *South Latitude,* 186-8, 339.
5 Douglas, *Rough Log,* 42.
6 Douglas, *Rough Log,* 45.
7 Douglas, *Rough Log,* 46.
8 Douglas, *Rough Log,* 30.
9 *Captain's Forty Ninth Letter.*
10 Deacon, Twentieth Scientific Report, 1936.
11 Deacon, 1939, 193.
12 NOL MS 2/4/117, Letter from Under Secretary of State, Colonial Office ref. 26169/36, February 14, 1936, 7.
13 *The London Gazette, 14th February 1936,* col.1, issue 34255, page 9722.
14 Douglas, *Rough Log,* 53.

Chapter 31: Melbourne Re-visited

1 Ommanney, *South Latitude,* 222; *Captain's Forty Ninth Letter.*
2 NOL MS 2 1/3 2379-238, Broadcast from Melbourne, February 17, 1936.
3 Hill to Nattriss, 5th April 1936.

4 NOL MS 2/3/460 (22690/2), Telegram to Master, *Discovery II,* 10th January 1936.
5 NOL MS 2/3/460 175, Telegram no 423 sent to Master, *Discovery II,* 12th February 1936.
6 NOL MS 2/3/460, Telegram from Captain, *Discovery II,* to Discovery Committee, 26th January 1936.
7 NLS MS 9560/127-137, Telegram from Hill, *Discovery II* to Secretary, Discovery Committee, 2nd March, 1936; Reply 25th April, 1936.
8 NLS MS 9560/136, Horsburgh to Discovery Commission, May 13th, 1936.
9 NOL MS 2/3/501, *Captain's Fiftieth Letter of Proceedings for R.R.S. Discovery II,* Fremantle, 7th April, 1936.

Chapter 32: Epilogue: A Normal Cruise

1 Hill and Douglas, *Report.*
2 SPRI MS 1284/5;CC, Discovery Committee, 150th Meeting, March 23rd, 1936.
3 Courtesy of Spinks 9136243. 33C CO 448/45 11891499.
4 NOL MS 2/3/460 (22690/29), Melbourne Herald to Captain, R.R.S. Discovery II, 11.2.36.
5 NOL MS 2/3/26277, Telegram 423, Master Discovery II to Discovery Committee, 9th February 1936.
6 Discovery Committee, 150th Meeting, March 23rd, 1936.
7 Douglas, Vice-Admiral Sir Percy, Mr Ellsworth and the Discovery II, *The Polar Record* no 12: July 1936.
8 Tribute by Mr H.G. Maurice following Deacon, 1939, 206-207.
9 NOL MS 2/3/459 64, Telegram from Discovery Committee to Master R.R.S. Discovery II, 30th December 1935; *Captain's Fiftieth Letter.*

Select Bibliography

Borchgrevink, C.E., *First on the Antarctic Continent: being an account of the British Antarctic Expedition*, London, George Newnes Limited, 1901.

Coleman-Cooke, John, *Discovery II in the Antarctic: The Story of British Research in the Southern Seas*, Odhams (Watford) Ltd, 1963.

Davis, J.K., *With the* 'Aurora' *in the Antarctic, 1911-1914,* London: Andrew Melrose, Ltd, Nabu Public Domain Reprints.

Haddelsey, Stephen, *Ice Captain: The Life and Times of J.R. Stenhouse*, The History Press, 2008, 132-133.

Hardy, Sir Alister, *Great Waters: A Voyage of Natural History to study whales, plankton and the waters of the Southern Ocean in the old Royal Research Ship,* Discovery, *with the results brought up to date by the findings of the R.R.S.* Discovery II, Collins, London, 1967.

Ross, M.J., *Ross in the Antarctic. The Voyages of James Clark Ross in Her Majesty's Ships Erebus and Terror 1839-1843*, Caedmon of Whitby, 1982.

Scott, Captain Robert F., *The Voyage of the Discovery,* first published 1905, Nonsuch edition, 2007.

www.ingramcontent.com/pod-product-compliance
Ingram Content Group UK Ltd.
Pitfield, Milton Keynes, MK11 3LW, UK
UKHW032104150325
456262UK00003B/324

9 781836 150374